AF884566

PEOPLE OF MANIPUR

PEOPLE OF MANIPUR

ANTHROPOGENETIC STUDY OF FOUR MANIPUR POPULATION GROUPS

RAMA CHAKRAVARTTI

1986

B. R. PUBLISHING CORPORATION
DELHI-110052

First Published 1986

ISBN 81-7018-296-4

© Author

Published in India by:

B.R. PUBLISHING CORPORATION
461, Vivekanand Nagar
DELHI-110052 (INDIA)

Distributed by:

D.K. PUBLISHERS' DISTRIBUTORS
1, Ansari Road, Darya Ganj
NEW DELHI-110002 (INDIA)
Phone : 2 7 8 3 6 8

All rights reserved with the publishers including the right to translate or to reproduce this book or parts thereof except for brief quotation in critical articles or reviews.

Printed in India at:

RENU PRINTERS
SHAHDARA, DELHI-110032

Dedicated to
Late Prime Minister
Indira Gandhi

Preface

This study covers a period of over 20 years of painstaking and labourious field investigation.

It was in 1960 that, during my first sojourn at Manipur, I had an opportunity to visit five districts of this small hilly state. I will ever remember with sincerest gratitude the kind help and cooperation of the Tribal Welfare Department of the Government of Manipur and especially of late Duncan, the then Officer-in-Charge of the Department at Imphal. He would have been very pleased to see this publication, had he been alive for a few more years! Although most of those concerned with the documents, involving a great number of friends, admirers and well-wishers belonging to the Government of Manipur, non-Government and semi-Government organizations are presently lost, yet their memories are still fresh in my mind.

I had another opportunity to visit my old contacts, and those still surviving most gladly extended their help and cooperation in my study. I express my gratitude to Messers Schweizerbartsche Verlagsbuchhandlung, Stuttgart, West Germany, for kindly permitting the reprinting of the article that appeared in their journal. I owe a deep sense of gratitude to all those who have extended their generous help and valuable cooperation for the completion of this study.

RAMA CHAKRAVARTTI

Contents

List of Plates

Introduction

The sources of history of Manipur, both antiquarian and recent, are as varied and extensive as those of any other region of India, the land of diverse ancient civilizations and culture. There are positive evidences to show that a fascinating civilization took its shape in the valley of Manipur several thousand years ago, and its further proliferation and diversification was sustained by the dynamic socio-political, economic and religious activities of the succeeding generations.

Despite the availability of rich and extensive literature of Manipur, there seems to be hardly any bio-anthropological account of the population groups of Manipur drawn on the basis of adequate scientific data. It was only during the year 1978-79 that a research project on Biology of Man in Manipur, was undertaken by the author under the auspices of Jawaharlal Nehru University Post-graduate Centre, Imphal. Funds for the project were provided partly by the Tribal Welfare Department of the Government of Manipur.

The progress made so far on anthropogenetic study of the peoples of Manipur is thus far from satisfactory, and a good deal of spade work is still needed to obtain dependable results in this direction.

The history of Manipur can be divided into three phases: (i) the first phase relates to the period when the entire population lived in hilly areas and the compartmentalization between hill and valley areas had not happened; (ii) the second one relates to the period when the migration of people from the hills to end their habitation of the valley commenced, and (iii) the third phase covers the period when the distinctions between the peoples in the hill and those in the valley were markedly concertized and led to socio-economic and political interactions.

Relying upon the Manipuri legends and historical records, it can be concluded that the ancient forefathers of the Manipuris had their origin in the hill areas of Manipur. This is testified by the presence on the border between Manipur and Burma of the place, Chingnunghat, which literally means a hole through the interior of a solid hillock.

Tombi Singh (1972) has made a scholarly study of the culture, geography, political history and economic potentials of the state of Manipur. The ethnic homogeneity of the population groups of the hill and the valley population groups has been emphasized in his following account (1972 : 17):

> "Many outsiders including the learned section think that there is a basic difference between the valley people of Manipur and those who are in the hills. In fact this is not so. The entire people of Manipur belong to the same ethnic group and trace their origin more or less to the Sino-Tibetan group of human species."

However, at present the gap between the valley people and hill people seems to have considerably widened. Two factors have mainly contributed to such cleavage: firstly, the difference between language and facilities of communication and agriculture. The second factor accentuating the distinction is the conversion of the major population of the valley into Hinduism nearly four centuries back, which ushered in a new dawn of orthodoxy and conversatism in the valley. Hinduism still continues to be the principal faith of the Maitis inhabiting the valley. Despite such distinctive features, the interaction between the hill and valley continues to remain cordial and co-operative. To quote again Tombi Singh (1972 : 17):

> The virtues of tolerance and understanding bind the two

major sections of the population inspite of their various differences.

Lately, the differences between the hill and the valley people appear to have been accelerated further. An attempt has been made in this book to bring into sharp focus the biogenetical similarities as well as dissimilarities of the major population groups of Manipur. We have, however, refrained from passing any remarks on the superiority of an ethnic group, since no evaluation is admissible in a biological project.

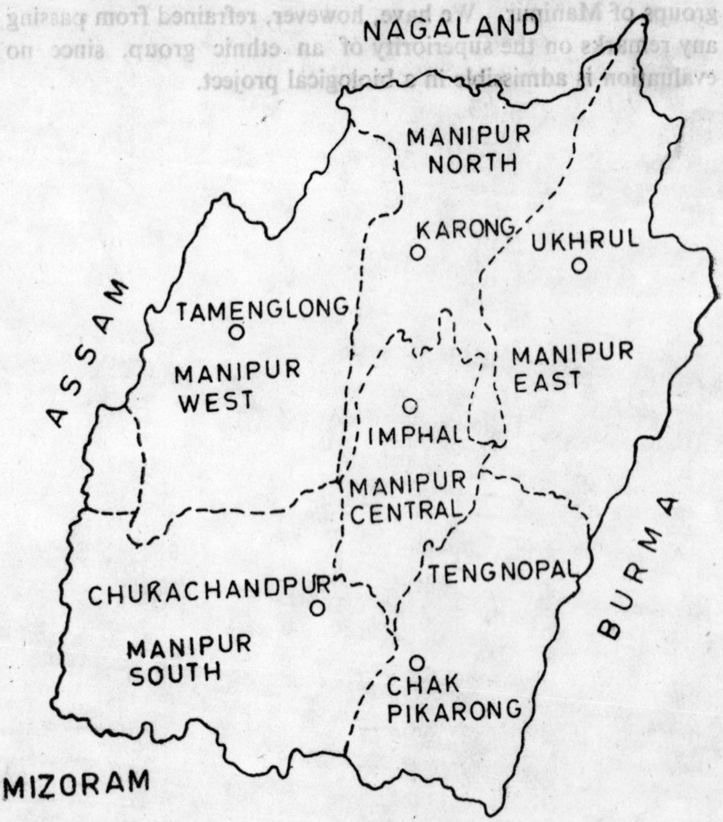

MAP OF MANIPUR

1

Land and People of Manipur

Few areas of the world can rival with Manipur in the rich diversity of its population, culture and above all its panoramic view. Engulfed by adjoining areas of Nagaland, Assam and Tripura, Manipur is the homeland of bewildering variety of peoples and kindreds, tribes and tongues. In old days, Manipur was known to her neighbouring states by different names given by them. In Rennal's *Memoir and Maps of India*, it was called "Mecklay". In the narrative of *symes* and in Maps of that period Manipur was called "Cassy". To the Shans it was known as "Kase" and to the Burmese as "Kathe", a corruption of the same word. The Ahoms called it "Makeli" and the Cacharis "Magli"; while the old Assamese name for it was "Moglan". There are differences of opinion whether the name Manipur is associated with this land from very ancient time or in recent period. Whatever it may be, it can be safely asserted that the name Manipur of this land became popular only in the modern age.

Manipur is bounded on the north by Nagaland; on the north-east and south by Burma; on the south-west by the union territory of Mizoram and on the north-west and west by Assam (Fig. I). It extends from latitude 23.83°N to latitude 25.68°N. Its westernmost tip is 93.03°E longitude and eastern most one is 94.78°E longitude. In terms of area it is not a big state. Manipur covers an area of 8,628 square miles (approximately 22,356 square kilometre), of which 700 sq. miles constitute the valley where the Manipuris or Meiteis live. The rest is the homeland of various hill tribes. The population of Manipur state as per 1971 census record 10,73,758; of this 7,10,169 are the Meiteis. They constitute 70% of the total population of the central valley. The religion-wise break-up of the population of Manipur, according to 1971 census, is as follows:

Population groups:		% *of total Population*
1. Hindus	6,32,697	58.97
2. Christians	2,79,243	26.03
3. Muslims	70,969	6.62
4. Others	89,944	8.38

The state of Manipur has five districts:

(a) Central district (b) West district (c) North district
(d) East district (e) South district

Manipur is a hilly region. About 10% of the area is plain and the rest is covered by hills. Most of the plain is covered by Manipur valley which is about 1,920 square kilometre in area and is about 750 metres above the sea level. The remaining portion of the plain is in the Jiribam subdivision on the Manipur side of the Assam-Manipur border. The great Indo-Gangetic plain bifurcates into two valleys near the western border of Meghalya. One is the Brahmaputra valley and the other, the Surma valley. The latter covers Sylhet district of Bangladesh and ends at the western fringe of Manipur. The Manipur valley is centrally located. Imphal, the capital of Manipur, is in the valley.

Manipur has been a part and parcel of India from dim antiquity. Recensions of the Mahabharata prevalent in northern India refer to Manipur in four different places. In some recensions prevalent in other parts of the country it is variously referred to as "Manphur", "Manlur" or "Manyur". Manipur was one of the land routes on the eastern frontier of India through which Indian culture passed into upper Burma. Likewise, Manipur was one of the gates through which Mongoloid tribes, their languages and beliefs poured into eastern India. The Himalayan ranges on the north of India are lofty, snowbound and impassable except at some places. So they succeeded to a large extent in segregating the races living on their northern and southern sides. Indo-Aryans and others to the south of Himalayas and Mongoloid races to the north of them have been living for thousands of years, comparatively unmixed with one another. But the hills separating India from Burma are lower and less impassable. So mixing of races took place much more extensively and much more thoroughly. Manipur was the melting pot on the eastern gate of India where races, languages and cultures met through ages, wave after wave, influencing one another to a large extent but ultimately mingling together.

Although a part of India culturally and geographically, Manipur remained aloof from the main political trends of the country. None of the great ancient empires, such as the Maurya Empire, the Gupta Empire and Harshavardhana's Empire reached Manipur. The Muslims conquered most parts of India but they could not subdue Manipur. Even the Mughal Empire at its zenith did not include Manipur. However, during British days, there was no direct British rule over Manipur and it was allowed to remain as a native state with all degrees of aloofness.

For a description of the physical features, the state may be divided into three regions:

(i) the central valley; (ii) the hills surrounding the valley; and (iii) jiribam subdivision. The ranges of hills surrounding the valley are in a sense offshoots of the Himalayas. The Shangpo or the Brahmaputra after flowing in a west-to-east direction on the north of the Himalayas makes a sharp bend and enters Assam near 95° E longitude. Ranges of mountains still continue from this point towards the east. From the latter ranges, there are offshoots first running in a south-westerly direction forming the hills of Nagaland, and then running in a southernly direction forming the hills of Manipur.

The ranges of hills running in the north-south direction bifurcate in the northern part of Manipur, a series running to the west of the valley and other series running to the west of the valley. These two series meet again in the southern part of Manipur, thus enclosing the Manipur valley. About two third of the population of Manipur live in this valley. The ranges of hills of Manipur, are higher in the north, reaching well over

2,700 metre above the sea level and then gradually diminishing in height as they run towards the south. Not only the hills but also the valley gradually diminishes in its height as one proceeds towards the south. For this reason all principal rivers of the state flow from north to south. The southern part of the valley is studded with lakes of different sizes. There is a belief that the entire valley was once a big lake and that it became a dry, habitable land by silting up. Innumerable rills and streams and the new rivers bring down silt from the hills every year and make the valley fertile and enable it to support a dense population.

The state has a wide range of climate. The climate of the valley is temperate and delightful. The valley is served by the monsoon but the rainfall is not very heavy. The annual rainfall in the valley is about 140 cm. This rain is mostly from the south-west monsoon. The climate in the north-eastern hills is colder and somewhat wetter. The annual rainfall at Ukhrul is about 170 cm.

The most important rivers of Manipur are the Imphal, the Iril, the Thoubal, the Irang and the Barak. All these rivers flow from north to south. The lowest portions of the state are in the southern half of the valley. The important lakes are Logtag, Ikop, Pumlen and Waithon. Logtag is the biggest of them all. All the lakes are important as fisheries.

During the years 1881-1901, there was an increase of population by 63 thousands, i.e., average annual increase by 3.1 thousand in Manipur. But the sharp increase and population explosion could be noticed during the last 70 years, between 1901 and 1971, with an average annual increase of 7.8 thousand population. It is estimated that during the last 70 years there was almost 25% increase in the total population within this state.

Four categories of people have mingled together in the formation of the Manipuri-speaking people and they are:

(a) people living in the valley before A.D. 33;
(b) people colonising the central part of modern Imphal west in A.D. 33;
(c) immigrants from other parts of India;
(d) immigrants from upper Burma.

When Pakhangaba arrived in Imphal in A.D. 33 with fire and sword, the valley was sparsely populated. In consolidating his principality, he thoroughly exterminated his enemies. Those who did not like to face liquidation, either fled to the cold hills of the north-east or the north or surrendered to him and accepted his authority. Those who fled to the cold hills mostly perished in the cold climate while some merged with the hill population. Those who surrendered to him intermarried with the colonisers and thus became Manipuri speaking people. Pakhangaba seems to be an Indian prince seeking his fortune on the Indo-Burma border. In the first and the second centuries of the Christian era, there were many such princes.

Pakhangaba's successors annexed, in the course of centuries, the other principalities in the valley and all the hills around it and extended their territory in the Kabaw valley and beyond the upper course of the Chindwin called the *Ningthee* by them. All those settlers who came by about the close of the 19th century were absorbed in the local population, and went into the formation of the Manipuri-speaking people. The Shans who came to Manipur during 15-17th century probably merged among the Manipuris. The immigration from the rest of India continued upto the end of the 19th century and is still continuing. The miscegenation briefly delineated above will explain the great diversity of facial features and the difference in the stature of the various people in Manipur.

A little less than one third of the population of the state belongs to what are called scheduled tribes. To pay special attention to their development, there is a department called the Tribal Welfare Department which has been functioning since 1955. The scheduled caste population is small—it is about 2% of the entire population.

THE SONS OF THE HILLS: I

In 1908, Hodson stated that "Meithei" has been derived from Mi=Man and Theie=separate. "In the moietay of Manipur we have the combined appellations of the Siamese Tai and the Kochin Chinese Moy. In other words, the Manipurian tribe, called Cossiah by the Bengalis belongs to the Moi section of the great tribe called the Tai by themselves and Shanvel Syan by the Burmese, the sectional name being also foreign and equivalent to the native."

Chatterjee (1946) in his famous "Kirata-Jana-Kirti" stated "the meitheis or Manipuris are the most advanced section of the Kuki-Chin people". "They have their kinsmen in Burma and appear to have settled in ancient time in Manipur and the Lushai hills, as well as in the Chittagong hill tracts. These Indo-Mongoloids are known to the Assamese as Chins and the Kuki-Chin has been adopted as a composite and inconclusive name for them. Singh (1975) described the Meitheis as:

"The people of Manipur in the valley, speaking as their mother-tongue, the Manipuri language, is known in the local language as the Meiteis. Whenever references have been made about them in any book, the word Manipuri has been used to denote the Meiteis."

The Manipuri language is also known as "Meitei Loi" which means the language of the Meiteis. The Meiteis are broadly divided into Hindus, Muslims and scheduled castes. The Meitei Muslims or Manipuri Muslims have many things in common with their Hindu counterparts, in respect of language, dress and construction of huts. The caste rules in Manipur do not appear to be very rigid and differ very much from those in Bengal. The modern Manipuri race is a composite one formed out of several tribes. The fertile valley of Manipur witnessed the invasions of different tribes from time immemorial. At different periods, the Nagas, the Kukis, the Shans, and the Chinese came and settled in this land and merged themselves into the Manipuri community.

The majority of the Manipuris call themselves "Kshatriyas" and wear the sacred thread. Among the Kshatriyas, there is a community known as Bishnupriya Manipuris. They are said to be "the descendants of 120 Hindu families of different castes, who were brought into the valley of Garib Niwaz in the later half of the 18th century to teach the indigenous inhabitants of the valley the custom of the Hindus. They intermarried with the people of the country but after a time the Meitei or the original Manipuris came to the conclusion that the new comers were of inferior stock" (Roy, 1958). According to another version the Bishnupriyas are supposed to have been the first ruling race, and the Meiteis are supposed to have been the next immigrants. Their language is distinctly different, the Khalachai (Bishnupriya) language is more akin to the Bodo—Chinese group. It is quite a reasonable surmise that the Austric Kha-Chais (Khasis) who had submitted to the influence of Bodo and the Asura culture and stayed in the plains of Kamrupa. When that country was over-run by those races, they gradually turned into a mixed race with a mixed culture and language. At a later time, due to fresh political turmoil in Kamrup, instead of migrating to the Khasis and the Jaintia Hills where they were socially banned by the orthodox batch of their race, they migrated further eastwards and settled in the valley of the wide Loktak lake or Khala and even known as Khalachais (Roy: 1958). The headquarter of these people were at Vishnupur, so named after the tutelary deity Vishnu. The Bishnupriyas claim the status of bonafide Manipuris of their formation of section of the Manipuri population. This has to be examined with emphatic references to their position in Manipur and the affinity of the language and other ethnic factors. They have a dialect absolutely strange to the Manipuri language in structure and in original character. The emergence of this community in the context of the recent history of Manipur must have been consequent upon the contacts between the kings of Manipur and other Indian communities, especially of the eastern zone, comprising Assam and West Bengal. Their dialect looks like more deformed Bengali than Manipuri. The proof of Bishnupriya's relation with Manipuri in the recent past, is however, given by the vocabulary contents of the dialect. The dialect contains quite a large number of Manipuri vocabularies. Among the Kshatriyas there is a class known as Rajkumari. They are the descendants of the various Rajas of Manipur. A Rajkumar is not allowed to marry a Rajkumari. The title of Maharaj Kumar is borne only by the sons of the ruling chief. The Brahmins enjoy a very high position in Manipur. The ancestors of the Brahmin family came and settled at different times in Manipur mostly from Bengal. At present they are endogamous, although formerly they used to intermarry with the Kshatriyas. There is a class of writers. They do not wear sacred thread and still retain the Bengali titles. They are known as Lairikyengbum. They still retain titles like Basu and Das. But in other respects they have now been absorbed into the general mass of the Manipuri population.

The Loi is not recognised as a pure Manipuri. They appear to be descendants of the former inhabitants of Moirang, one of the original tribes which formerly occupied the valley to the south. They are mostly engaged in salt making, silk manufacturing and fishing.

The ancestors of the Muslims of Manipur came at different times from Sylhet and Cachar. They took Manipuris as their wives and settled in Manipur. They have adopted Manipuri as their mother-tongue but in dress and customs quite often they maintain the strict Islamic standards. The population which left Manipur in the exodus towards the west, also included Manipuri Hindus (the Bishnupriyas) and the Manipuri Muslims. Thus one can see big pockets of Manipuri Muslims living in close neighbourhood with Manipuri Hindus as members of the same family in Cachar, Tripura and Sylhet.

Ethnically, culturally and linguistically the Meithei is an organised and consolidated community having distinctive traits of a sub-nation. They can clearly be distinguished from the people living in the hills.

Ethnological Affinities of the Manipuris

W. McCulloch (1859)

"From the most credible traditions, the valley appears originally to have been occupied by several tribes, the principal of which were named Koomul, Looang, Moirang and Meithei, all of whom came from different directions. For a time, the Koomul appears to have been most powerful, and after its declension, the Moirang tribe. But by degrees the Meithei subdued the whole, and the name Meithei has become applicable to all. Since their conversion to Hindooism, the Meitheis have claimed for themselves a Hindoo descent."

Pemberton (1835)

"We may safely conclude them to be descendants from a Tartar colony from China."

For this conclusion, Hodson (1908) stated:

"I can see no reason and think there is far more ground to conclude them to be descendants of the surrounding hill tribes. Tradition brings the Moirang tribe from the south, the direction of the Kookies, the Koomul from the east, the direction of the Murrings, and the Meithei and Looang from the north-west, the direction of the Koupoees. The languages of the Murrings, Kookies and Koupoees, are all very similar, and as the Koomul, etc., the offshoots of these tribes were, as said before, at different periods the dominant tribes in the valley, it might be expected that the present language of the people, united under the name of Meithei, would have a very apparent likeness to these languages, and such is the case. All these tribes have also traditions amongst themselves that the Munnipories are offshoots from them."

Presently the upper classes of Manipuris deny their origin from the hill tribes surrounding the valley, although, when asked to account for themselves otherwise, they have no plausible story to offer. They are of the opinion, that they always belonged to the valley and always have been a separate race.

BROWN (1874)

"Although the general facial characteristics of the Munnipories are of the Mongolian type, there is a great diversity of feature among them, some of them showing a regularity approaching the Aryan type. Among both men and women, the stature is very various, differing about as much as is found among Europeans. Some of them are very good looking and fair. It is not uncommon to meet with girls with brownish-black hair, brown eyes, fair complexions, straight noses, and rosy cheeks. The Munnipories are decidedly a muscular race, some of the men particularly so; they are generally spare in habit of body, and fat people are rare. They have good chests and well-formed limbs. The men are not heavy, though averaging about five feet seven inches. The women are four inches shorter than the men."

"The Lois very closely resemble the Meitheis, and are to all intents and purpose indistinguishable from their over-lords. Indeed, it is remarkable to observe how much the individuality of such people as the Meitheis, the Lois, the hill people and even the Gurkhas, depends on differences of clothing and coiffure.

In the case of the Lois, who are either earlier settlers or the direct descendants of Meitheis banished to Loi villages as a punishment, this phenomenon is intelligible, for the infusion of true Meithei blood into Loi families by mixed marriages is undeniable."

"The Panggan or Muhammedan settlers are distinct from the Meithei, as is reasonable to expect, although it has been noted that many of the Muhammedan inhabitants of Sylhet and Cachar resemble the hill type. The Panggans are believed to have originated from Cachar as prisoners of war taken by Meitheis."

About the origin of the Manipuris, there exists two excellent versions by Brown (1874) and Hodson (1908).

BROWN (1874):

"By the end of Dapar Jug and beginning of the Kali Jug (year 3435) Enoog Howba Chonoo, the wife of Babroobahan gave birth to a son called Pakhangba. He used to assume the form of God by the day and by the night he used to be a man."

Hodson (1908) collected the following genealogy:

"The Brahma Dev (the creating God of the universe) had sprung out from the navel lotus of Narayana (the protecting God of the universe). Marichi Muni (a Hindu sage) was born from the limbs of Brahma Dev. Marichi's son, Kosshop Muni (a Hindu saint); Kosshop Muni's son, Surja (the Sun); Surja's son, Shaborna Muni 1; Chitraketoo; Chitroketoo's son, Chitradhaja; Chitradhaja's son, Chitrabija; Chitrabija's son Chitrasarba 1; Chitrasarba's son, Chitra Rat; Chitra Rat's son, Chitra Vanoo; Chitra Vanoo had no son, only a daughter named Chitrangada; Chitranagada's son, Babrubahan; Babrubahan's son, Soopra-bahoo; Sooprabahoo's son, Pakhangba (jobista). The Jobista or Pkhangba was the first ruling king of Manipur."

Ethnological Notes on the Meitheis (*Waddel*: 1901)

MITAI OR MANIPURI

"This Indo-Chinese tribe is fast becoming Hinduised into a caste, claiming to be Kshatriyas or Rajputs, though its members are not admitted to be such by orthodox Hindus. In keeping with their new position, their chief is now called a Raja and a Brahmin priest has invented for him a pedegree tracing his descent to Arjun, the hero of the Mahabharat. Still even now he cannot be installed as a Raja till he and his wife go through a ceremony in which they are clad in the costume of the surrounding savage Naga and Kuki tribes. His house is built on the pattern of a Naga hut and a man armed with a Naga spear and shield accompanies him as a guard."

"That their veneer of Hinduism is still very thin as evidenced by the recent massacre of Europeans at the Raja's village. They occupy the rich alluvial valley known by the Hindu title of "Manipur" a tributary of the Chadwin branch of the Irawadi, and so are outside the Brahmaputra basin though politically at present within Assam. But several of the tribes have settled in the latter area of recent years as traders, and this is where I got most of my specimens for measurement, in a settlement in the Sibsagar district."

"Physically the men are fairly tall and muscular. The men dress as Hindus and wear the Brahmanical thread. On account of their posing as Hindus they would not allow me to measure their facial angle, the instrument which is designed to go between the closed teeth. They protested that their caste would be broken, were they to take into their mouth any object which had been in the mouth of another person of a different caste. The women dress generally like Hindu females and also cover their head with their shawl. The unmarried girls crop their hair over the forehead to form a short fringe as seen in the annexed photograph. Wives are purchased and seem to be the slaves of the husband; when in debt, he occasionally sells them.

"Their affinities appear to be with the western Nagas and Kukis, though their rather fair complexion and better features suggest relationship with the Shans, and in the chronicles of the Pong Shans it is recorded that a Shan army visited this valley in A.D. 777. The Burmese call their country "Ka-the" (that is Ka-se)."

Waddel (1901) measured 15 individuals from five areas. Their anthropolometric measurements are given below. The 15 Manipuris were sampled from the following areas:

(a) Manipur valley : 7
(b) Sibsagar : 3
(c) Sylhet : 2
(d) Nowgong : 2
(e) Dibrugarh : 1

Mean Somatometric Measurements (all in mean):

1. Stature	=	1636 mm
2. Weight	=	128 pounds
3. Span	=	1699 mm
4. Sitting Ht.	=	828 mm
5. Head Length	=	184 mm
6. Head Breadth	=	142 mm
7. Min. Frontal	=	100 mm
8. Byzygomatic	=	133 mm
9. Bigonial	=	100 mm
10. Nasal Ht.	=	47 mm
11. Nasal Breadth	=	36 mm
12. Nasal Projection	=	28 mm
13. Naso-malar breadth	=	113 mm
14. Bimalar breadth	=	99 mm
15. Vertex to chin	=	221 mm
16. Vertex to nasal notch	=	105 mm
17. Lt. Humerus	=	309 mm
18. Lt. Calf Girth	=	33 mm
19. Lt. Radius	=	258 mm
20. Lt. Middle Finger to Middle of patella	=	133 mm

Colour of the Iris: No. 3

Meitei Villages: Imphal East Block: Manipur

Rural Development Plan: Association of Voluntary Agencies for Rural Development (AVARD): (1975)

The survey undertaken by AVARD in 16 Meitei villages presents a peculiar situation. Although most households (98 per cent) comprise small and marginal farmers, owning less than two hectares of unirrigated land, yet there is no starvation or dire poverty. The secret is that most farmers are peasant proprietors outside the clutches of money lenders and they grow their own food grains. They are not exploited by big landlords who in the process twist the agriculture system to produce for the export market where buyers/traders may have the upper hand, or by outside traders who directly force unfavourable terms upon them. The subject area is almost self-reliant. About ten per cent of the local produce, at the present pattern of consumption, makes the area surplus, and the imported goods make about equivalent component in the household expenditure statement.

Another notable feature of the area is its high participation rate, especially among women. About 80 per cent of the women in the age-group 16 to 60 are either earners or earning dependants. They help their husbands and/or parents in farming, selling of fish and other items, and more commonly produce handloom cloth. This high female participation rate should have given the area a higher per capita income, but the productivity of an average worker is low. The reasons are: absence of sufficient productive assets and ignorance about improved methods of production. The prime problem of development, therefore, is of raising productivity of an average worker so that the situation where people work but still remain poor is improved.

Physically the state is well-knit, isolated unit in the extreme north-eastern corner of India. Although as of today the area cannot be classified as inaccessible, still its integration with the rest of the country continues to be slow. The easiest approach to Manipur is by air. And air service inspite of possible subsidies, would continue to be costlier than rail/road transport. Manipur also suffers from the disadvantages of being a border state. Again the state has to contend with the special problems of a sizeable tribal population of Nagas and Kukis who till late were left aloof from the mainstream of national life.

The state is broadly divided into two natural regions, the valley and the surrounding hills. The valley is largely inhabited by Meiteis and the hills by Nagas and other tribes. The valley, a natural rice bowl, is thickly populated. As opposed to it, the hillfolk pursue primitive subsistence agriculture and their areas are sparsely populated. Between the hills and the valley are foothills.

It is difficult to classify the population of Manipur by economic conditions. The average holding does not exceed one hectare and about two-thirds of them are below the average. No wonder less than half (48 per cent) of the population is below the poverty line by per capita income of Rs. 600/-per annum at current prices. The proximity to Imphal has one other effect. Whatever little savings of the people, those are either spent on imported goods or invested in the town. The result: not much investment has been taking place to improve local agriculture. The crop yields here are lower than at many places in the valley.

The literacy rate here is not very low (22 per cent). It is higher in the villages nearer the town and decreases as one goes farther. The area has no mines or minerals, excepting a little special type clay in Andro village (AVARD, 1975). Excepting the ubiquitous weaving among Meitei women, there is no other skill worth reporting.

Topographically it is, by and large, plain but most villages extend upto the foothills. As a result, some parts of their land are slopy and undulating. Some villages like Thiyamkonjin, Bashikhongba and Khanarok have their cultivated land on hill slopes. The slope is from north to south. The Iril river provides the main drainage. The valley is 2,560 ft. above mean sea level. Its geology is still unexplored. Due to altitudes summer is not severe. Winter, though cold, is not extreme. The climate is on the whole healthy. The average annual rainfall in the valley area is 129,63 cm. The months of June-August are the wettest months.

THE SONS OF THE HILLS: II

In the hills, we find different tribes such as the Tangkhuls, the Maos, the Kabuis, the Marrings and a number of allied tribes such as Thadous, the Paites and the Gangtes. They constitute about 31% of the total population of the state. The people living in the hills are all Mongoloid in appearance. But the tribes differ very much among themselves in height and body-build. The Marrings living in the hills in the south-eastern part of the state are of slender built with medium height. The Tangkhuls living in the north-eastern hills are tall and usually slender. The Maos living in the northern extremity of the state are usually stoutly built. The Kabuis inhabiting the western and the north-western hills are usually of slight build. The Kukis live in the south-western hills in small settlements.

The extraordinary heterogeneity of the hill peoples of Manipur makes it difficult to give a thorough account of the customs, traditions and legends of the hill peoples as a whole. There is a legend among the hill peoples which says that the Nagas, the Kukis and the Meiteis are descendants from a common ancestor, who had three sons. The Nagas are descendants of the eldest son, the Kukis are descendants of the second while the Meiteis are the descendants of the youngest. The tradition varies in detail from hill tribe to hill tribe.

Hodson (1908) described: "While accepting with due reserve the familiar distinction between Nagas and Kukis, it may be pointed out that the tribes commonly classed together as Naga and Kuki occupy definite areas in these hills. A line drawn following the Kubo valley road via Aimole and joined to the Cachar road which traverses the western hills from Bishnupur boundary of the state separates the Naga area from the Kuki area without excluding more than a few small Kabui villages which lie to the South of Nongba. North of that line we find Naga villages and to the South of it Kuki settlements, although for various reasons, economic and political, Kuki settlements exist not only in the British district of the Naga hills in the vicinity of Henema but also far away on the north-east of the Manipur state, in the neighbourhood of Melome and Lapoome where the frontier has been delimitated and ascertained only within the last few years. Indeed the Kuki is to be found almost every where in the state except in the territory occupied by the Mao confederacy."

Differences of stature, dress, coiffure and weapons make it easy to distinguish between the members of these tribes. In colour they are all yellow-brown but with little variety, though some of the Tangkhuls who earn their living by salt making seem to be darker. Among them all, as among the Manipuris, there are persons who have a tinge of colour in their cheeks when still young. The nose also varies, for there are cases where it is almost straight, while in the majority of individuals it is flattened at the nostril. No information is at present available as to the measurements of the skull.

In the case of the Tangkhuls, the traditions as to their origin resolve themselves into at least three groups. In the first group are the traditions which connect them with the valley and derive them thence as emigrants on account of the heat and mosquitoes. According to the second group of traditions, the village of Hundung is the centre of their dispersion and the stories declare that they sprang from immigrants from the village of Maikel Tunggam. The third set of traditions refers to a period rather lately and mainly deals with the growth of new villages and their establishment by emigrants from older villages which suffered from an excess of population. The second group of tradition affords a link with the general common ancestor, who had three sons who became the progenitors of those tribes. This tradition, which is widely spread, agrees in its many versions in assigning the primacy of descent to the Kukis, the next place being to the Nagas, while the Manipuris are said to be the children of the youngest of these three brothers.

BROWN (1874)

"they say they came out of a cave in the earth at a place called Murringphy in the hills, about 4 days' journey north-east of the Munniporie valley."

As regards the Kabui Nagas, McCulloch (1859) reported that: "Songboo tradition gives, as the place of their origin, the mountain towards the south of the valley named, Thungching. They state themselves to be at present located on the sites of villages at one time occupied by the southern tribes, who are, they say, the elder branch of the family of which they themselves are the middle and the Munniporees the younger branch." He further added: "the Munniporees place their origin at Kayboching, near the Aquce route, north of the Government Road, whence they spread to the South."

The fact is that all these legends are aetiological, invented to explain a state of affairs which was recognised as needing some sort of an explanation. Geographically, the hill areas of Manipur are contiguous to Burma in the east and south-east, Assam in the west and Mizoram territory in the south-west. The land routes connecting Manipur valley with Burma, Cachar district and Nagaland lie across these hills. But very little intercommunication existed inside the hilly region, between the hills and the valley as well as between the hills and the neighbouring states. This is responsible for the isolation and the material backwardness of the people living in this area. Otherwise, the entire area composed of olive green hill features under the deep blue sky is quite attractive having temperate and healthy climate. Some of the places like Ukhrul, Mao, etc., may be developed into beautiful hill resorts. The hill area is rich in forest and some mineral resources like lime and slate. Tamenglong area is suitable for tea plantation. Lime and plentiful bamboos may easily feed sizeable cement and paper industries. The sons of the hills consist of 29 tribes. Among these, the Tangkhuls, the Thadous the Kabuis, the Maos, the Paites, the Hmar and the Kacha Naga form numerically sizeable tribes. Ethnically these 29 tribes of Manipur may broadly be classified into:

A. The NAGAS, and
B. The NON-NAGAS (the Kuki-Chin groups)

The tribes of the Kuki-Chin family are not eager to form a sub-nation. They prefer to be identified individually by names of the respective tribes.

The Kukis arrived late in Manipur (in the period between 1830 and 1840). The Kabuis, the Tangkhuls and the Marrings have lived in contact with the plain people for many centuries. The Marrings live in the south-eastern part of the state. The Tangkhuls living in the north-eastern hills and the Kabuis in the western and the north-western hills and in small settlements throughout the valley. The Kukis live in the south-western hills of the state.

THE A GA TRIBES

The Tangkhul, the Roungmei (Kabui), the Mao, the Kacha (Liangmei), the Maram, the Marring, the Anal, the Monsang, the Lamgang, the Moyon, the Zemi and the Thangal are the important Naga tribes of Manipur. Besides these, there are some Angami and Sema Nagas. The Zemis and Kabuis are allied tribes. They live in the Tamenglong subdivision in the western hills of the state. The Maos, Maram and the Thangals live in the northern subdivision of Mao and Sadar hills bordering Nagaland. The Tangkhuls live in the Ukhrul subdivision adjacent to Burma. The Marrings, the Moyon, the Monsang, the Lamgang and the Anals occupy the Tengnopaul subdivision in the south-western part bordering Burma. Originally, the Moyon, the Monsong, the Lamlang and the Anals belonged to the Old Kuki Clans. But they have changed their affiliation and are now included in the Naga sub-nation. These tribes in addition to their Kuki-Chin traits have assimilated some Naga characteristics in their ways of life. Indeed, they form a bridge between the Naga and the Kuki-Chin tribes of Manipur.

THE KUKI-CHIN TRIBES

The Aimol, the Chothe, the Chiru, the Kom, the Purum, the Gangte, the Paite, the Simte, the Thadou, the Vaiphei, the Hmar, the Zou, the Ralte, and many Mizos are the non-Naga tribes of Manipur. Ethnologically, they belong to the Kuki-Chin family. These tribes live in the Sadar hills, Churachandpur, Jiribam, and the Tengnopaul subdivisions of Manipur. According to Shakespeare (1912) the so-called old Kuki clans, the Aimol, the Chothe, the Chiru, the Koireng, the Kom, the Purum, the Anal, the Lamgang, the Moyon, the Monsang, the Gangte, and the Vaiphei came to Manipur from the hills south of Manipur in Burma in the 16th century. The so-called new Kuki clans, the Simte, the Paite, the Zou and the Hmar, etc., came to Manipur in A.D. 18th and 19th century. Large scale immigration of the Kuki tribes in Manipur in the 19th century is on record. The Paite is a group of people in the process of migration.

Of the two tribal groups of Manipur, the Naga tribes have ethnic and cultural affinities with the people of Nagaland and the non-Naga tribes belonging to the Kuki-Chin family have affinity with the Chins of Burma and the Mizos of Mizoram.

All these tribes have their kith and kin scattered in Cachar, Assam valley and in upper Burma. Many tribal villages have sprang up in the Imphal valley also. The recent trend among the educated and well-to-do people belonging to various Naga tribes is to own houses in Imphal to enjoy amenities of modern city life. The Kuki-Chin tribes have their maximum concentration in Chura-chandpur which is fast developing into a modern township and in order of importance second only to Imphal, the capital of the state. The life of the hillman is very much influenced by their surrounding environment. They set up their villages on the top or slopes of high hills covered with forests. In selecting sites for their villages they must have been guided by two considerations: the security and health. Frequent inter-tribal conflicts made them security conscious. Hill tops are not only easily accessible but they also allow distant views which help in detecting enemy movements from far off. But the Kuki-Chin groups preferred the foot of the hills and the rich valleys in the hills. Search for hunt and fertile soils have still kept them on the move.

There are numerous dialects among the hill people of Manipur. In Ukhrul for instance, each village has its own dialect. In the Linguistic Survey of India, these dialects have been broadly classified on the basis of their affinity into two groups: the Naga groups consisting of Thangal, Maram, Kabui, Kacha and Tangkhul and the rest are included in the Kuki-Chin family. There are indeed great similarities among the Kabui, Thadou, Mizo and Meitei. The Tangkhuls can pick up Meitei at ease. Most of the non-Nagas understand each other through Thadou. None of these tribal dialects has any script. Among the various tribes in Manipur, the sole medium of intercommunication is Manipuri. From time immemorial it is the lingua-franca of the state. Indeed the Manipuri language has brought unity and understanding among the various linguistic groups in Manipur.

APPENDIX

WADDEL (1901): MANIPURI NAGAS

The Manipuri Nagas lie outside the Brahmaputra valley, on the Burma side. Of these the chief are the "Luhupa", so-called by the Manipuris on account of the helmet-like crest of hair which they have along their crown. They never wear metal earrings, only plugs of wood, and the men wear glans rings as described by Brown (1874). They build rest-seats paved with stone in memory of departed chiefs, like the Himalayan tribes and corresponding to the Zayats of the Burmese. Kolyas and the Tangkhuls on the north and west are rather degenerated Nagas. According to Waddel (1901), the Manipur Nagas are classified as below:

A. Western Nagas

(a) *Kabui* (b) *Koirang* (c) *Sengmai* (d) *Tangkhul*
e) *Marring*

B. Central Nagas

(a) *Kolya* (b) *Maram* (c) *Mao* (d) *Luhupa*
(e) *Miyang Khang*

Anthropometric Measurements

There are no measurements of the Tangkhul Nagas by Waddle (1901). Only one Kabui Naga was measured by him. The measurements are mentioned below:

1. Height=1674 mm 2. Head Length=179 mm 3. Head Breadth=147 mm 4. Minimum Frontal=99 mm 5. Bizygomatic Breadth=138 mm 6. Nasal Height=46 mm 7. Nasal Width=35 mm 8. Nasal Projection=29 mm 9. Nasc-Malar Breadth=122 mm 10. Bimalar Breadth=101 mm 11. Lt. Calf Girth=34 mm 12. Vertex-Nasal notch=111 mm 13. Vertex-Chin Length=230 mm 14. Weight=95 pounds

Iris Colour : No. 2

Informations Related to the Scheduled Tribes of Manipur

I. Percentile distribution of workers and non-workers among the scheduled tribes:

	Workers		*Non-workers*	
	Males	*Females*	*Males*	*Females*
Total	48.33	52.12	51.67	47.88
Rural	48.60	52.68	51.40	47.32
Urban	35.08	11.80	64.92	88.20

II. All scheduled tribes who are registered as rural in Manipur:

Total	244,903
Males	120,753
Females	124,150

		Rural	*Urban*
Kabui Nagas	*Total*	26,883	2335
	Males	13,384	1229
	Females	13,499	1106
Tangkhul Nagas	*Total*	43,585	358
	Males	21,346	225
	Females	22,239	133

III. Marital status and age for the tribes:

Kabui Nagas:	*Total*	29,218	*M*:	14613	*F*:	14605
	Unmarried:	*M*: 9400	*F*:	8258		
	Married:	*M*: 4820	*F*:	4959		
	Widowed:	*M*: 274	*F*:	1063		
	Divorced and Separated:	*M*: 56	*F*:	243		
Tangkhul Nagas:	*Total*	43,943	*M*:	21571	*F*:	22372
	Unmarried:	*M*:13724	*F*:	13065		
	Married:	*M*: 7597	*F*:	8686		
	Widowed:	*M*: 224	*F*:	548		
	Divorced and Separated:	*M*: 21	*F*:	53		

IV. Age groups

	Male	*Female*
Kabui Nagas		
Age 0-14 years	*6823*	*6181*
Not married	6816	6170
Married	7	11
Age 15-44 years	*6065*	*6190*
Not married	2468	2013
Married	3476	3704
Widowed	55	29
Divorced and Separated	21	137
Age 45 +	*1651*	*2188*
Not married	64	44
Married	1316	1232
Widowed	218	793
Divorced and Separated	35	105
Tangkhul Nagas		
Total	*21571*	*22372*
Not married	13724	13065
Married	7597	8686
Widowed	224	548
Divorced and Separated	21	53
0-14 years	*9986*	*9849*
Never married	9986	9833
Married	—	16
15-44 years	*8763*	*10,110*
Never married	3666	3206
Married	5048	6667
Widowed	40	204
Divorced and Separated	9	33
45 and above	*2783*	*2404*
Never married	45	22
Married	2537	1999
Widowed	184	343
Divorced and Separated	12	20

V. Demographic Aspects

The annual rate of increase of the population of Manipur is 3.5 per cent and as for the sex-ratio, there are 1015 females per 1000 males. The density of population is about 35 persons per square kilometre. The density of population in the valley area of Manipur is higher. The average density in the hill areas is only 16 persons per square kilometre.

There is a downward trend in sex-ratio during the decade 1951 to 1961. The average increase in respect of Manipur as a whole during the intercensal period from 1951 to 1961 is 35 per cent. As against the decennial growth-rate of 21.5 per cent for India as a whole, the increases shown by Manipur (also Assam and Tripura) are very high. This is due to the excess of births over deaths during the intercensal period though immigration is also a factor.

The level of fertility also must have contributed to the high growth-rate. Nuptial fertility rate for Manipur is not available, but if the high rate prevailing in Assam be of any indication, it may be assumed that the nuptial fertility rate in Manipur is one of the highest in India.

The nuptial fertility rate per 1000 women aged 15-44 in Assam is 317.0, the highest in India. The state proportion of women aged 14-44 to total population vary within the range of 20 to 22 per cent. Manipur has returned an unusually high

Percentage of age-groups 0-14 and 15-49 and of married women (15-49) to total population and children (0-14) to married women (15-49) in Manipur.

Age group 0-14			*Age group 15-49*		
Persons	*Males*	*Females*	*Persons*	*Males*	*Females*
43.12	21.75	21.37	51.08	25.12	25.96

Percentage of married women (15-49) to total population	Percentage of children (0·14) to total married women (15-49)
16.34	95.42

proportion of 25.96 per cent of women within the age group of 15-49 to the total population.

It may safely be concluded, therefore, that the high nuptial fertility rate and the high proportion of women of the reproductive age-group to the total population must have contributed substantially to the high growth rate of population.

It will be seen from the above that the proportion of males in the age-group of 0-14 is greater than that of females, while in the age-group of 15-49, the proportion of females is greater than that of males. The percentage of married women of the age-group of 15-49 to the total population is 16.34.

Children of the age group 0-14 account for 95.42 per cent of the total number of married women of the age group 15-49.

Manipur has a comparatively high mean decennial growth-rate of 29.82 per cent which is greater than that of the neighbouring major state of Assam (29.39). The mean decennial growth rate for India as a whole is 19.52 per cent.

Manipur has barely one to one-half per cent of its population as urban, the remaining is rural.

A feature of the India sex-ratio that causes some concern is the widening excess of males over females in almost all parts of the country. But there are exceptions. Manipur, Kerala, Goa, Daman & Diu, Pondicherry, Laccadiv islands have excess of females over males. For all India the number of females per 1000 males is 941.

Age and sex-ratio :

0-14		*15-49*		*50+*		*Age not stated*	
M	*F*	*M*	*F*	*M*	*F*	*M*	*F*
43.84	42.42	44.84	45.79	11.11	11,57	0.21	0.22

Females per 1000 males
1015

The 1961 sex-ratio classified by age-groups does not show uniform trends in the three subdivisions where the present research investigation has been conducted. First of all, we will examine the total females per 1000 males in the age-groups in the three subdivisions: Imphal East, Imphal West and Ukhrul. This will be compared against the figure of Manipur.

Subdivisions:	*0-04*	*5-9*	*10-19*	*20-29*	*30-39*
Imphal East	1001	996	1013	1100	1019
Imphal West	1045	964	964	980	972
Ukhrul	988	953	997	1250	1006

Subdivisions:	*40-49*	*50-59*	*60-69*	*70 & above*
Imphal East	1014	1065	1028	1332
Imphal West	1065	1113	1238	1429
Ukhrul	956	1032	1099	1282

The subdivision of Ukhrul has a constant deficiency between age-groups of 0 and 19 as well as 40-49. Imphal East is deficient in the age-group of 5-9 only while Imphal West is deficient between age-groups of 5 and 39 years.

It may, however, be observed that in Manipur as a whole the number of females per 1000 males, belonging to age-groups of 5-9, 10-19, 30-39 and 40-49 is deficient while in the other age-groups there is generally an excess of females over males.*

The proportion of widowed and divorced or separated females per 1000 males, in each age-group is very large. This is so even in the age-group of 10-14. The conclusion that may be drawn from this is that widowed and divorced or separated females do not generally get married whereas in the case of their male counterparts, remarriage takes place after they become widowers or are divorced or separated.

Manipur has been making a rapid, almost spectacular progress in literacy during the last 30 years. There is a progressive

*The corresponding figure of Manipur:

0-4	5-09	10-19	20-29	30-39	40-49	50-59
1010	975	993	1112	961	996	1007

60-60	70 & above
1075	1210

rise among the males though females have yet to go a long way to catch up with their male counterparts.

The percentage of literacy in Imphal East and Imphal West are upto 25 per cent above the average rural literacy. Lack of communication facilities is the main cause of the comparatively low rate of literacy returned from the hill subdivision of Ukhrul although there is reasonable Missionary influence. Imphal has easy communication facilities and comparatively good distribution of educational institutions. The distribution pattern of workers as cultivators, agricultural labourers or at household industry who have also secondary work, which is obtained in Manipur does not conform to the overall pattern in India. Nor is the pattern similar to that obtaining in the neighbouring state of Assam. A feature that attracts notice is that a large proportion of females who work principally as cultivators is also engaged at household industry and even greater proportion of females whose principle work is household industry also work as cultivators.

In the hill areas which constitute the greater part of Manipur, extensive cultivation is not possible because of the terrain and indifferent quality of the soil. As if to make up for this deficiency, nature has provided the little valley with a very fertile soil: in fact, the Manipur valley may be regarded as one of the most fertile areas in the North-Eastern part of India.

Natural Vegetation and Forest Resources

Geologically, the country is of recent origin. The upper tertiary continues through the Mapum and Poshing to Kopamedza range in Angami Nagaland. Kasom range and Mungshongkhong are of cretaceous age. The valley has alluvium soils (recent). The soil at the base of the hills has its origin from sand stone rocks, the upper layer of which contains deep humus. At Langal, it is ferrugious clay to clayey foam. Forest resources play a very important role in the economy of the state. Out of the total area of the state, 64.2 per cent are covered with forests. Out of the total area of forests, reserved forest constitutes 9.3% while protected forests are 28.7%. Forests in the state are divided into four zones:

(1) *Burma border forests*
These forests lie on the Indo-Burma border.

(2) *Ukhrul (north-east) forests*
These are scattered all over the hills of Ukhrul (east) district.

(3) *Forest overlooking the valley*
These forests have no specific name. They are named after the hills on which or the villages near which they are situated.

(4) *Jiri-Barak drainage forests*
These forests are situated in the hills to the west of the valley along the course of the Barak river and its tributaries, e.g., Jiri, Irong, Leimatak, Mukry and other side streams on the Manipur Cachar border.

Our knowledge of the various plants of Manipur is limited. About 200 species of flowering plants and pteridophytes were described till date. The natural vegetation of Manipur is severely affected due to human interference. Most of the forests show useless secondary formations. Almost all the hills around the valley are represented by a mat vegetation of grasses with associated herbs and shrubs. Some of the main causes of the deforestation are discussed below.

(a) Fire

Jhum cultivation is practised in many parts of the hill areas in the state. There being no control on the fire, it spreads through the vegetation of the different ranges displaying a grand "Diwali" festival at night for weeks together until a heavy shower of rain puts them off during the months of March and April. The intensity of the burning is well experienced in the valley when a pretty thick layer of cloudy smoke covers the sky and causes respiratory and allied discomforts to every one. On many occasions, the fire is caused by people without any reason.

(b) Timber and fuel wood

Merciless and non-planned cutting of woods for timber and

fuel is one of the major causes of quick disposal of natural vegetation in the accessible forests near the human habitations. Initially, the larger species of trees are felled, followed by smaller species of trees and shrubs.

(c) Network of roads

Development of roads is progressing rapidly since the second five year plan. In the beginning of the second five year plan, there was only 960 km of roads in the state which increased to 1705 km in 1961, 2400 km in 1966 and 2600 km in 1971. At present, this has definitely increased further. Many of these roads pass through the dense, previously inaccessible and undisturbed natural forest ecosystems of the state. These networks of roads are the symbols of prosperity but, on the other hand they help people consciously or unconsciously to exploitation of natural vegetations rapidly and thereafter quick transport of the exploited materials by high power diesel vehicles.

(d) Land reclamation

Many areas in the hills and valley have been reclaimed either for cultivation or human inhabitation at the cost of natural vegetation due to increased population.

(e) Intruder plants

The two important species of intruder plants and their role in the destruction of natural vegetation could not be ruled out. These are, Lantana Camara and Mikamia Micrantha. Lantana came to Manipur probably 30-35 years ago. Previously, it was limited to some areas of the valley. Now it could be seen in various interior areas of the state. This is a very aggressive invader and grows fast in disturbed vegetation. In a short period it grows thick and does not allow any other plants to grow with it. The growth has now slowed down and has steadily began to invade the vegetation nearby. Mikamia is a gregarious climber and probably it came to Manipur hardly eight or ten years ago. Its devastating activity is much more than that of the Lantana. It has already arrived at various hilly areas of Manipur. It can withstand all the extremes of climate conditions. Rapid growth of the plant forms a dense thick cover over any plant or vegetation thereby almost stops the photosynthesis of the hosts causing their death within a short period of time.

In some of the hill areas the vegetations have been disturbed for security purposes. The autotrophes are the only organism capable of trapping solar energy by the process of photosynthesis and absorbing essential mineral and chemical substances from a biotic environment so that the energy and nutrients flow through various food chains to the different strata of heterotrophic organisms comprising animals, bacteria, fungii and other micro-organisms. Some of the immediate effects of the destruction of the natural vegetations are worth mentioning.

The only source of water here is the rain as there is no snow clad mountain in the state. Plant cover helps in absorbing maximum rain-water by the soil during the monsoon months and retaining enough of this so that streams, springs or rivers may get a remarkable flow of water during the dry months of the year. In many parts of the hills the people suffer severely due to acute scarcity of water as the source of water dry up during summer months. The heavy downpours fall directly upon the upper humus layer of the soil. There being no plant cover to protect the soil from the impact of these torrential drops, penetration of water in the soil takes place to a greater extent and initiates the soil erosion primarily with the removal of the upper organic nutrient layer along the slopes of the hills. Subsequently, the inner layers of soil are also drained out ultimately exposing the hard stony layer. These eroded areas affect the nearby vegetations leading to frequent land slides.

Undertaking of research activities in Manipur is highly warranted especially on the following lines.

(*i*) *Floristics*

There are many regions in Manipur which has hardly been explored by any biologist till date. Eastern part of Manipur gets the benefit of migration of plants from the adjoining Burmese flora. There are many such records of plants which have enriched the flora of India. The Manipur Lily (Lilium Macklinae) from Eastern hills and the two species from Manipur namely, Gleadevia banerjinese and Hoya manipurensis have thus been discovered. The discovery of Saparia Himalayana from Manipur connects missing link of distribution of the family from Malayan archipelago to Khasi, Abor and Mishmi hills. Identifying of isoetes in Manipur has been placed on record for the first time in the history of north-eastern India.

(ii) *Prevention of soil erosion*

Detailed plan with local expertise would be helpful in preventing soil erosion.

(iii) *Studies on timber yielding plants of Manipur*

There are a number of timber yielding plants in Manipur. But scientific data of their distribution, natural regeneration, etc., are lacking. Exotic plants can easily be introduced. Manipur has introduced a number of exotic plants which have responded satisfactorily in the local environment.

(iv) *Essential oil yielding plants of Manipur*

There are many essential oil yielding plants growing wild in Manipur. Scientific research will reveal information so that there is every possibility of inclusion of many of these plants to the list of aromatic plants of economic importance. Agar plants grows scattered in various part of Manipur. All the plants are not economically important as the presence of essence is limited to a few plants due to infection of a fungus. Generally the method of obtaining Agar practised here is a trial and error method. Proper scientific investigations will invariably help in inducing infection to a higher percentage of Agar plants.

(v) *Medicinal plants of Manipur*

Our knowledge of medicinal plants in Manipur is incomplete and meagre. Biochemical investigation would definitely help in finding out the economic and pharmaceutical values of many such plants.

(vi) *Wild life conservation*

Wild life is a part of the forest complex. The rapid deplition in the recent years of the stocks of wild life is a matter for serious concern. Brown antler deer locally known as "Sangai" is one of the rarest and localised sub-species in the world. A thorough and intensive ecological survey in this context will save this sub-species from extinction.

(*vii*) *Introduction of plants for horticultural purpose.*

(*viii*) *New host plants for rearing "Tasar" and other types of silk worms.*

Unlike in other parts of the country, the value of realisation and concept of utility are entirely missing in Manipur. Schemes are, therefore, to be implemented to raise public awareness and to put them in the minds of the local talents.

References

1. Avard (1975) : "Meitei Villages: Imphal East Block, Manipur" *Rural Development Plan*, New Delhi.
2. Brown, R. 1874 : *Statistical Account of Manipur*, London.
3. Chatterjee, S.K. (1946) "Kirata-Jana-Kirti : The Indo-Mongoloids : Their contribution to the History and Culture of India", *J. Asia Soc., Bengal* P., 16 : 2.
4. Hodson, T.C. (1908) : *The Meitheis*, Reprinted by B.R. Publishing Corpn., Delhi, 1974.
5. *The Naga Tribes of Manipur*, Reprinted by B.R. Publishing Corpn., Delhi, 1974.
6. McCulloch W. (1859) : *Account of the valley of Manipur and of the hill tribes*, with a comparative vocabulary of the Manipur and other languages, Calcutta.
7. Perberton, R.B. (1835) : *Report on the Eastern Frontier of British India*, Calcutta.
8. Roy, J. (1958) : *History of Manipur*, Eastlight Book House, Calcutta.
9. Shakespeare, J., (1912) : *Lushei-Kuki Clans*, Macmillan, London.
10. Singh, R.K. Jhalajit (1975) : *Manipur*, Publication Division. Ministry of Information and Broadcasting. Government of India, New Delhi.
11. Waddel, I.A., (1901) : *The tribes of the Brahmaputra valley*, Reprinted by Sanskaran Prakashak, Delhi, 1975.

Additional Readings

12. Grimwood, E.S.C., (1891) *My three years in Manipur*. London.
13. Johnstone, J., (1896) : *My experiences in Manipur and Naga Hills*. Reprinted by Vivek Publishing House, Delhi 1971.

2

Reproductive Life of Manipuri Women

The study on the reproductive life of the women of four population groups of Manipur was done with the assistance of local research investigators. The main purpose of the study was to delineate a faithful account of fertility behaviour among the population groups studied.

DATA FOR INVESTIGATION

The Data for the study comprise:

(a) Meitei Manipuri: (192) villages: Thoubal, Singjamai and Lamphelpat

(b) Kabui Naga: (240) villages: Majorkhul and Andro

(c) Tangkhul Naga: (179) villages: Ukhrul and its neighbourhood

(d) Muslims or Pangan: (251) villages: Canchipur and Lillong.

Results and Discussions

Tables 2.0—2.3 report the age-wise distribution of ever married women, number of living and dead children, miscarriages and abortions alongwith the total number of pregnancies of the four studied groups.

TABLE 2.0

Age-wise Fertility Behaviour among the Four Population Groups

Population	*Age Group*	*No. of Living Children M*	*F*	*No. of Dead Children*	*Abortion*	*Total Pregnancy*
Meitei :	15-29 yrs. (59)	71	73	16	13	173
	%	1.20	1.23	0.27	0.22	2.92
	30-44 yrs. (85)	186	188	36	35	445
	%	2.19	2.21	0.42	0.41	5.23
	45 yrs. + (48)	105	104	55	10	274
	%	2.19	2.17	1.15	0.21	5.71
Total	192	362 189	365 1.90	107 0.56	58 0.30	892 4.65

TABLE 2.1

Age-wise Fertility Behaviour among the Four Population Groups

Tangkhul Naga:	15-29 yrs. (47)	49	58	7	2	116
	%	1.04	1.23	0.15	0.04	2.46
	30-44 yrs. (92)	205	192	21	11	429
	%	2.23	2.09	0.22	0.12	4.66
	45 yrs. + (40)	121	124	19	10	272
	%	3.03	3.10	0.41	0.25	6.80
Total	179	375 2.09	374 2.09	47 0.26	23 0.13	817 4.56

A. MEITEI MANIPUR (TABLE 2.0)

In the lower age-group (15-19 years), the mean fertility rate is 2.92, out of which the survival rate is 2.43 and the child-loss (total) rate is 0.49. In the middle age-group (30-44 years), the fertility rate is almost twice than that of the lower-age group (5.23). The survival rate and the total child-loss rates are 4.40 and 0.83 respectively. The upper age-group (45 years and above) shows only a slight increase in the total pregnancy (5.71) than that of the middle age-group. The survival rate is slightly lower with increased child loss rate (1.36). The abortion rate is more noticed in the middle age-group (0.41) which equals to the combined value of the remaining two groups.

B. TANGKHUL NAGA (TABLE 2.1)

The Tangkhul Nagas present an almost different picture from those of earlier two communities. The total pregnancy steadily presents an increased value from lower to upper age-group. The survival rate, total child-loss rate and abortion rate also agree with the trend of total conception rate, which increases with the advancement of age.

C. KABUI NAGA (TABLE 2.2)

There appears to be a gradual rise of total pregnancy value from the lower to the upper age group. The middle age-group shows more than double the rate of total conception. The survival rate behaves almost like that of the previous population group. The child-loss rate is appreciably high among the upper age-group. Similar trend is noticed in respect of abortion rate.

D. MUSLIM (TABLE 2.3)

The Muslims again follow the trend very close to that of the Tangkhul Naga in respect of conception, survival, child-loss and abortion rate. There is an appreciable increase in the death rate of the children in the middle and upper age-groups.

Table 2.4 exhibits the informations related to conceptions in the four population groups. There is practically no male-female sex-ratio difference among the four groups. The rate of abortion and miscarriages are more among the Meitei Manipuris,

TABLE 2.2

Age-wise Fertility Behaviour among the Four Population Groups

Population	*Age group*	*No. of living Children* *M*	*F*	*No. of Dead Children*	*Abortion*	*Total Pregnancy*
Kabui Naga :	15-29 yrs. (84)	91	85	25	8	209
	%	1.08	1.01	0.29	0.09	2.47
	30-44 yrs. (87)	235	212	74	5	526
	%	2.70	2.44	0.85	0.06	6.05
	45 yrs. + (69)	156	155	124	10	445
	%	2.26	2.25	1.79	0.15	6.45
Total	240	482	452	223	23	1180
		2.01	1.89	0.93	0.10	4.92

TABLE 2.3

Age-wise Fertility Behaviour among the Four Population Groups

Muslims :	15-29 yrs. (111)	134	149	49	11	343
	%	1.21	1.34	0.44	0.10	3.09
	30-44 yrs. (92)	243	228	115	17	603
	%	2.64	2.48	1.25	0.18	6.55
	45 yrs. + (48)	126	120	68	9	323
	%	2.63	2.50	1.42	0.17	6.72
Total	251	503	497	232	37	1269
		2.00	1.98	0.92	0.15	5.00

which is almost equal to the sum total of the remaining three groups. The Kabui Nagas have the lowest abortion rate. Table 2.5 presents the mean number of living and dead children along-with mean fertility rate among the four groups. The chi-square test on living vs. dead children reveals significant values ($P>0.001$), especially between the Meitei and Tangkhul; the Kabui and Tangkhul; the Tangkhul and Muslims as well as between the Muslims and combined Nagas of the hills as will be evident from the Table. The mean fertility rate is more among the Muslims while the three other groups behave almost uniformly.

Tables 2.6 and 2.7 describe the age-wise distribution of the onset of age at menarche as well as the menopause among the four ethnic groups of Manipur. Compared to the data on the age at menarche, the data on menopause is very scanty. Even then some interesting results emerge out. It may, however, be mentioned that the exact estimation of age among the groups studied was very difficult and we had to rely on their assessment of age, although among the two Naga groups many would produce birth certificates because of their affiliation to the churches.

Table 2.8 explains the mean age at menarche and menopause with test of significance between the groups. It will be evident from the Table that the Kabui Nagas record the highest mean age at menarche (15.15 years) while the Muslims have the lowest value (19.93 years). The t-test yields significant difference between the Meitei and the Muslims ($P>0.001$); Kabui and Muslims; Tangkhul and the Muslims. So far as the age at menopause is concerned, there is some significant difference between the Kabui and the Muslims ($P>0.02$) as expected. Much larger data with correct estimation of age may throw further light in this connection. Table 2.9 presents some data on the age at menarche among the North-Eastern Indian population groups. Excepting the Muslims of Assam, all the other caste groups present an uniform mean age at menarche, although the data from Assam is inadequate. The only data on the Zemi Nagas (Nagaland) presents a higher value than the mean menarcheal age of all Assam groups (combined). In this study, the Manipuri Muslims as contrast to Assam Muslims,

TABLE 2.4

Details of Conception among the Four Population Groups

Population	*Living children* *M*	*F*	*M/F*	*Dead Children*	*Abortion and Misc.*	*Total conception*
Kabui Naga (240)	482	452	1.06	223	23	1180
%	51.60	48.90		0.93	0.10	4.92
Muslims (251)	503	497	1.01	232	37	1269
%	50.30	49.70		0.92	0.15	5.00
Meitei (192)	362	365	0.99	1.7	58	892
%	49.79	50.21		0.56	0.30	4.65
Tangkhul Naga (179)	375	374	1.00	47	23	817
%	50.06	49.94		0.26	0.13	4.56

TABLE 2.5

Mean Fertility Rate in Four Population Groups

Population	*Mean living*	*Mean Dead*		*Mean Fertility*	$X^2 1$
1. *Kabui Naga*	3.90	1.02	4.92	*Kabui* X *Muslim*	0.04
2. *Muslim*	3.98	1.07	5.00	*Meitei* X *Muslim*	2.33
3. *Meitei*	3.79	0.86	4.65	*Meitei* X *Kabui*	1.78
4. Tangkhul Naga	4.18	0.38	4.56	*Meitei* X *Tangkhul* =34.88 ++	
				Kabui X *Tangkhul* =54.15 ++	
				Tangkhul X *Muslim* =58.63 ++	
				Muslim X *Meitei* =15.47 ++	

++ P> 0.001

TABLE 2.6

Age-wise Distribution of Menarche among the Four Groups

Age Class	*Meitei*	*Kabui*	*Tangkhul*	*Muslim*
7 -9 yrs.	—	—	—	4
%	—	—	—	0 .99
10-11 yrs.	—	1	3	37
%	—	0.21	0.98	9.20
12-13 yrs.	65	4	93	130
%	13.35	0.83	30.29	32.34
14-15 yrs.	283	211	122	164
%	58.11	43.59	39.74	40.79
15-16 yrs.	132	251	89	64
%	27.10	51.86	28.99	15.93
17-18 yrs.	6	17	—	2
%	1.23	3.51	—	0.49
19 yrs. +	1	—	—	1
%	0.21	—	—	0.26
Total	487	484	307	402

have a very early menarcheal age. The Meiteis behave like the Zemi Nagas, while the Kabui and Tangkhul, the two Manipuri Naga groups present almost different picture. The importance of socio-economic and many other factors affecting the menarcheal age has been stressed by a number of researchers (Chakravartti, 1974). Peller (1967) stated "sexual maturation is an automatic endogenous process but environmental factors speed it up or delay it by months and years." If the environmental factors are sufficiently changed to influence growth,

TABLE 2.7

Age-wise Distribution of Menopause among the Four Groups

Age Class	*Meitei*	*Kabui*	*Tangkhul*	*Muslim*
26-31 yrs.	4	—	—	—
%	3.25	—	—	—
36-41 yrs.	36	—	12	60
%	29.04	—	17.65	60.00
41-46 yrs.	24	16	20	32
%	19.35	14.29	29.41	32.00
46-51 yrs.	20	72	32	8
%	16.12	64.29	47.06	8.00
51 yrs. +	40	24	4	—
%	32.26	21.42	5.88	—
Total	124	112	68	100

TABLE 2.8

Mean Age at Menarche and Menopause

Population	*Menarche*	*t values*	*Menopause*	*t values*
Meitei	14.34 yrs.	*Meitei* X *Kabui* =0.92	45.60 yrs.	0.85
Kabui	15.15 yrs.	*Meitei* X *Tangkhul* =0.52	48.86 yrs.	0.01
Tangkhul	13.93 yrs.	*Kabui* X *Tangkhul* =1.51	45.56 yrs.	0.59
Muslim	10.73 yrs.	*Meitei* X *Muslim* =4.51 ++ *Kabui* X *Muslim* 5.46 ++ *Tangkhul* X *Muslim* =4.51 ++	40.90 yrs.	1.26 2.32

++ P > 0.001

TABLE 2.9

Age at Menarche in North-East Indian Population Groups

Population	*n*	*Mean Menarcheal Age*	*Source*
Assam			
1. Kalita	172	12.61	*Das* and *Das, 1967* (cf. *Chattopadhyay & Khullar 1969)*
2. Brahmins	27	12.51	-do-
3. Vaishyas	35	12.97	-do-
4. Kaibartas	30	12.89	-do-
5. Brahmins	60	12.46	-do-
6. Kayasthas	32	12.46	*Rakshit, 1960*
7. Kalitas	17	12.12	-do-
8. Muslims	29	13.35	-do-
9. Brahmin	26	11.85	*Shrivastava & Goswami, 1968*
10. Kayastha	11	12.00	-do-
11. Kalita	20	11.65	-do-
12. Ahom	50	11.52	-do-
13. Others	59	11.70	-do-
14. Ahom	164	12.96	*Gogoi, 1972*
15. Mikirs	52	12.75	-do-
Arunachal Pradesh			
1. Adis	109	14.45	*Duarah, 1969*
2. Singphos	95	12.59	*Kar & Mahanti, 1975*
Nagaland			
1. Zemi Nagas	214	14.13	*Bhowmick et. al., 1971*
Manipur			
1. Meitei	487	14.34	*Present study*
2. Kabui Naga	484	15.15	-do-
3. Tangkhul Naga	307	13.93	-do-
4. Muslims	402	10.73	-do-

menarche may change considerably. At present, girls mature much earlier and their menarche is much less spread over the age years, than was a century ago. This is due to the improved living conditions. With the progress of civilisation, the menarcheal age is becoming lower than it used to be in the past (1951).

The future studies on the age at menarche, may be planned, based on various socio-economic factors which are of significance from the view of moral and psychic environment. Reliable estimate of the age is also equally important in this connection.

Menopause or loss of reproductive capacity is not a sudden process in women. The time or period depends upon socio-economic factors and health culture. Reuban (1955) stated that better economic conditions and good food delay the age at menopause. Kamat and Kamat (1959) revealed that Indian women reach the menopause between 40 and 45 years : 42.2 years being the mean. The present study confirms the earlier statement of Kamat and Kamat (1959).

Agarwala (1967) reported that an Indian woman on an average gives birth to 6-8 children if her reproductive period is not disturbed. The Zemi Nagas (Bhowmick et. al., 1971) present the mean fertility rate as 4.3. Our study also agrees with Bhowmick and Bhowmick (1967) in this aspect excepting the Muslims.

It may, however, be mentioned that the available data on Muslims from Mysore (Agarwala, 1967), U.P. (Agarwala, 1967) also present very high fertility rate. The Muslims are found to have higher fertility than the Hindus. This is confirmed when we review the All India data on mean fertility (Chakravartti, 1974) where the significant difference exists between the Hindus and the Muslims in all states.

Majumdar (1960) reported the mean value of living children as 4.23 against the mean value of dead children as 1.66 in his U.P. survey. Our series cannot be compared with any other data due to non-availability of materials from the neighbouring groups.

Table 2.10 presents the rate of abortion when the data is arranged in low, middle and upper age-groups. While examining the abortion rate of the four ethnic groups, it appears that

TABLE 2.10

Abortion rate in Four Ethnic Groups of Manipur

Age-group	*Meitei*	*Tangkhul*	*Kabui*	*Muslim*
Low (19-29 yrs.)	13	2	8	11
Medium (30-44 yrs.)	35	11	5	17
Upper (45 yrs. & above)	10	10	10	9
Total	58	23	23	37
%	0.30	0.13	0.10	0.15

among the Tangkhuls the abortion rate is almost uniform in middle and upper age-groups. The Meiteis have significant abortion rate in the middle age-group while there is equally high trend in other two age-groups. The Muslims behave like the Meiteis, while the Kabui Nagas have high rates in the lower and upper age-groups.

SELECTION INTENSITY

It has been earlier (Spuhler) pointed out that the greater the mortality and the greater the variations in fertility, the greater is the opportunity for selection to operate. Studies on Indian population groups—Basu (1967): Ghosh (1970)—have earlier suggested that the opportunity for selection increased as the number of children born per family decreases, if the mortality rate is assumed to be the same. The variation in fertility is directly related to the total selection which acts upon a population. Crow (1958), however, stressed that the changes in the two components with a changed environment may show striking difference. In our study, the effect of demographic differences in four contemporary populations on the Index of Opportunity for selection has been experienced. These four groups, all living in a relatively small area, show at one time the kinds of changes that often occur secularly in the same population. The demographic changes that are being narrated here, not only change the opportunity for selection but also have had qualitative effects by changing the criteria for selection. The total selection on a population may be expressed by a quantity which Crow (1958) calls the Index of Total Selection (I).

$$I = Im + If/ps$$
$$Im = pd/ps$$
$$I_f = Vf/\overline{X}$$

Where $\overline{X}$ is the average number of live births per women aged 40 years and above; Vf is the variance in progeny number due to fertility; pd is premature deaths (died before the reproductive age); $ps = 1-pd$; Im is the mortality component and If is the fertility component. The index of total selection intensity measures the maximum potential rate of change by selection (where O indicates no change) (Livingstone and Spuhler, 1965).

The contribution of differential fertility (If) appear to increase as family size increases as observed by Crow (1958). Variation in infant mortality and variations in fertility both play a role. When we examine our data (Table 2.11) the following facts emerge :

(a) the $\overline{X}$ value varies among the four contemporary communities between 5.84 and 6.79, excepting the Meiteis, an uniformity in respect of mean value is observed. The lower value of the Meitei may be due to the lower value of If (0.026).

(b) The pd is 8.45% among the Tangkhuls while this is almost uniform with the three other groups. This reflects the lower rate of premature deaths among the Tangkhuls which may be due to their improved living conditions and habitation in an improved ecozone.

(c) the lower value of Im and If among the Tangkhuls again confirms the above statement.

(d) The Im is highest among the Muslims, closely followed by the Kabui Nagas and the Meiteis. The If is highest among the Tangkhuls and lowest among the Meiteis.

The overall results, however, do support Crow (1958) that the I increases with the decrease of the family size.

It will be seen from Table 2.11 that the three valley population groups namely, the Meitei, the Kabui and the Muslims show gradual increase of mean fertility values alongwith the increase of the I value. The almost similar trend among these three communities, occupying the similar area, and exposed to similar environment clearly supports Spuhler (1962), that I

TABLE 2.11

Selection Intensity among the Four Ethnic Groups of Manipur

Population	I_m	I_f	I	$\overline{X}$	V_f	p^d	p^s
Muslim	0.325	0.043	0.382	6.72	2.05	0.245	0.755
Kabui	0.323	0.035	0.347	6.45	1.46	0.244	0.756
Meitei	0.311	0.026	0.340	5.84	0.885	0.237	0.763
Tangkhul	0.093	0.041	0.138	6.79	1.99	0.085	0.915

remains relatively high in city and town populations. These three population groups inhabit the central valley at Imphal town. The lower value of I among the Tangkhul Nagas living in the hilly areas in the Eastern zone of Manipur is very interesting. The mortality component of them is lowest among the four groups studied while the fertility component is almost similar to those of the other three population groups. The p^d is again very low than those of the other three groups and the p^s is again high.

The Tangkhuls live in a different ecozone and they have a very different way of life than other groups in question. Differential fertility seems to keep the selection intensity high despite the relatively low pre-adult mortality. However, it is to be noted that the above formulation is based on the assumption that all variations in mortality and fertility have a genetic basis and fitness is completely heritable. The index provides, therefore, not the net intensity of total selection but only its upper limit. Further, the genetic evaluation for varying values of *If* may be quite different, depending upon whether the population studied uses birth-control measures or is still under natural conditions. Therefore, caution should be taken in the conclusion of the results.

References

1. Agarwala, S.N., (1967) *Some problems of India's population*, Asia Pub. House, Bombay.
2. Basu, A., (1967) "Selection intensifies in the Paharis", *Eug. Quart.* 14: 241-2.
3. Bhowmick, K.L. and A. Bhowmick, (1967) "Some aspects of reproductive life of the Zemi women in Nagaland", *Res. Bull. Inst. Soc. Studies*—Summer 2: Calcutta.
4. Bhowmick, K.L., M.K. Chowdhury; P. Das and K.K. Chaudhuri, (1971) "Fertility of Zemi women in Nagaland," *Inst. Soc. Studies*, Calcutta.
5. Chakravartty, Rama, (1974) *Differential Fertility in a village population of coastal Andhra Pradesh*, Ph. D. thesis (unpublished) submitted to Dibrugarh University, Dibrugarh, Assam.
6. Chattopadhyay, P.K. and S. Khullar, (1969) "The age at menarche in India", *Acta Med. Auxol.* 1 : 58-62.
7. Crow, J.F., (1958) "Some possibilities for measuring selection intensities in man", *Hum. Biol.* 30 : 1-13.
8. Das, P.B. and B.M.Das, (1967) "Age at menarche of Kalita girls of Assam", *Man in India*, 47 : 113-117.
9. Duarah, A. (1969) Cf. Kar, R.K. and S. Mahanta, (1975).
10. Ghosh, A.K. (1970).
11. Gogoi, D., (1972) "Menarche and Menopause among women of an Ahom village in upper Assam", *Bull. Deptt. Anth.*, Dibrugarh University, Dibrugarh, Assam.
12. Kamat, M. and R.G. Kamat, (1959) *Suggestions for the effective control of population India*, Bombay.
13. Livingstone, F.B. and J.N. Spuhler, (1965).
14. Majumdar, D.N., (1960) "Social Contours of an industrial city", *Social Survey of Kanpur*, 1954-60, Asia Publishing House, Bombay.
15. Peller (1967) "Menarche: Unsuspected relations", in *Quantitative Research in Human Biology and Medicine* John Wright & Sons, Bristol.
16. Rakshit, S., (1960) "A short note on menarcheal age of Assamese girls", *Man in India*, 40 : 52-55.

17. Reuben, S., (1955) "Statistical data of the Medical aspects of Menopause in Indian women", *J. Assocn. Med. Women India*, vol. 43.
18. Sarkar, S.S. (1951) "The place of human biology in Anthropology and its utility in the service of the nation", Presidential Address: *Indian Science Congress Association*, Bangalore.
19. Shrivastava, R.P. and M. Goswami, (1968) "The menarcheal age of Assamese girls", *Anthropol.* 15.
20. Spuhler, J., (1962) "Empirical studies on quantitative Human Genetics", in *Proceedings of UN/WHO Seminar on the Use of Vital and Health Statistics for Genetics and Radiation Studies*, United Nations, New York, pp. 241-252.

3

Anthropometry

The technique of anthropometry is used in the present study so as to study morphology and body composition of the population groups of Manipur. As stated earlier that our investigations on anthropometry was limited within the following four groups of Manipur.

I. Meitei (Hindus)
II. Pangan (Muslims)
III. Kabui Naga (Christian Tribal group from the hill)
IV. Tangkhul Naga (-do-)

Altogether seven anthropometric measurements have been taken on each adult subject (both sexes): These are :

(i) *Stature* (ii) *Head Length* (iii) *Head Breadth*
(iv) *Nose Length* (v) *Nose Breadth* (vi) *Facial Length*
(vii) *Total Facial Breadth.*

Measurements thus recorded were later analysed and their mean values as well as standard error were computed. Three indices were also available out of these measurements which were classified after Martin and Saller (1959). The data were available from the villages of Singjamai, Lamphelpat in the case of the Meitei; Canchipore and Lillong for the Pangans; Ukhrul for the Tangkhul and Majorkhul for the Kabui Nagas. The age ranges from 20 to 36 years for the males and 18 to 35 for the females. The present data were subjected to comparison with those of available data and in the absence of any other data of recent origin, we have made an attempt to compare them with the available data from the North Eastern region of the country. Alongwith our present series we have also quoted the available earlier data on the various Naga and Kuki groups of Nagaland as well as the populations from Manipur state. This has given us further opportunity to make a comparison between these series. Martin and Saller's (1959) methodology was usually followed while locating landmarks and taking measurements. All the measurements were taken carefully by a single investigator and at a specific time of the day so as to minimise the personal error. Considerable difficulties have been encountered in the field in making a subject submit to the anthropometric ordeal and other genetic investigations. It has been observed that the subjects were to co-operate with us almost for the whole day and accordingly we had to compensate their wages either in terms of cash or in kinds.

I. MEITEI MANIPURIS

A sample of 105 adult males and 82 adult females were subjected to anthropometric investigation. The mean values for four important morphometric characters are given below:

TABLE 3.0

Mean Values of Stature and the Three Indices

Sex	*No.*	*Stature*	*C.I.*	*N.I.*	*F.I.*
Males	105	1603.59±5.03	76.93±0.30	73.42±0.62	87.78±0.47
Females	82	1504.08±5.63	76.67±0.42	72.71±0.71	88.66±0.80

The frequency distribution of these four traits and classification according to Martin and Saller (1959) are presented in Tables 3.1—3.4.

Both sexes are characterised by short stature although one-third of the individuals possess below-medium stature. The frequency of tall stature is almost same in both sexes.

TABLE 3.1

Distribution of Stature and Classification

Classification	*Males Frequency*	%	*Females Frequency*	%
Very short	1	0.95	—	—
Short	50	47.61	44	53.66
Below medium	32	30.47	24	29.27
Medium	11	10.47	8	9.76
Above medium	6	5.71	2	2.44
Tall	5	4.76	4	4.88

TABLE 3.2

Distribution of Nasal Index

Classification	*Males*		*Females*	
	f	%	*f*	%
Leptorrhine	33	31.43	26	31.71
Mesorrhine	69	65.71	54	65.85
Chamaerrhine	3	2.86	2	2.44

Both sexes are characterised by mesorrhine. One-third of the population have leptorrhine while only 3% of them are having broad and flat nose.

Almost 90% of the males and 93% of females present dolicho-mesocephalic head form. The males possess more broad heads than those of females.

More than half of both sexes have meso-leptoprosopic faces. The hyperleptoprosopic type of faces are more in the females.

TABLE 3.3

Distribution of Cephalic Index

Classification	*Males* f	%	*Females* f	%
Dolichocephalic	42	40.00	34	41.46
Mesocephalic	52	49.52	42	51.22
Brachycephalic	11	10.48	6	7.32

TABLE 3.4

Distribution of Facial Index

Classification	*Males* f	%	*Females* f	%
Hypereuryoprosopy	2	1.90	2	2.50
Euryoprosopy	25	23.21	10	12.50
Mesoprosopy	31	29.52	24	30.00
Leptoprosopy	31	29.52	26	32.50
Hyperleptoprosopy	16	15.24	18	22.50

than in the males. To sum up, the Meiteis are characterised by short stature, mesorrhine, mesocephalic and leptoprosopic face. There exists only one reference on the physical measurements of the Manipuris by Waddel (1900). On the basis of 15 measurements from the five areas (Assam and Manipur) he recorded the following mean values.

(i) stature: 1636 mm

Our study records the mean stature as 1603.59 mm. It is,

therefore, 33 mm or about an inch in the total body height of the Manipuris.

II. MANIPURI MUSLIMS

A sample of 98 males and 100 females were measured in respect of four anthropometric characters.

TABLE 3.5

Mean Values of Four Important Physical Traits

Sex	*No.*	*Stature*	*C.I.*	*N.I.*	*F.I.*
Males	98	1493.12±7.11	78.44±0.39	68.57±0.84	86.75±0.61
Females	100	1480.61±6.86	76.78±0.48	72.90±0.96	86.00±0.68

Excepting in the mean values of facial index, the males possess higher mean values in respect of the three characters. The difference between the two sexes in respect of stature appears to be about 12 mm which is expected. It may be stated that among the Meiteis this difference was noticed in a greater degree, i.e., almost 100 mm. The females were about 10 cm or about 4 inches less in height than those of the males among the Meiteis. No measurements are available on the Muslims of North-East India and as such any attempt to compare the present mean values of the Muslims of Manipur is not desirable. Risley (1891) measured Muslims from East Bengal (presently Bangladesh) and recorded the following mean values:

Stature : 1634.0 mm
C.I. : 78.0
N.I. : 77.5

Compared to these values our data shows an entirely different picture. Excepting the C.I., the other two mean values in the present series show very low values. Similarly, the mean value of stature recorded by Datta (1939) on the North Bengal

Muslims also show a high figure (1619 mm). It is, therefore, observed that not only the Muslims of Manipur show very low values in respect of stature compared to the other three groups studied, but also the mean value of stature as recorded among the Muslims, appears to be extremely low among all the North-Eastern Indian population groups.

TABLE 3.6

Distribution of Stature among the Muslims

Classification	Males f	Males %	Females f	Females %
Short	54	55.10	58	59.18
Very short	—	—	2	2.04
Below medium	18	18.37	16	16.33
Medium	14	14.29	18	18.37
Above medium	10	10.20	2	2.04
Tall	2	2.04	2	2.04

TABLE 3.7

Distribution of Cephalic Index among the Muslims

Classification	Males f	Males %	Females f	Females %
Hyperdolichocephalic	4	4.08	6	6.00
Dolichocephalic	20	20.41	46	46.00
Mesocephalic	64	65.31	44	44.00
Brachycephalic	10	10.20	2	2.00
Hyperbrachycephalic	—	—	2	2.00

Among the males more than 55% and among the females more than 60% are found with short stature. The very frequency of tall stature in both sexes is worth mentioning.

TABLE 3.8

Distribution of Nasal Index in Muslims

Classification	*Males* *f*	*Males* %	*Females* *f*	*Females* %
Leptorrhine	58	59.18	30	30.00
Mesorrhine	40	40.82	64	64.00
Chamaerrhine	—	—	6	6.00

Among the males the highest frequency is noticed with mesocephalic head form, while in the females about 46% are dolichocephalic. The broad-headed types are more found in the males than those of the females. The cephalic indices present almost a similar value with those of Risley (1891) and Datta (1939).

TABLE 3.9

Distribution of Facial Index among the Muslims

Classification	*Males* *f*	*Males* %	*Females* *f*	*Females* %
Hypereuryoprosopic	—	—	2	2.00
Euryoprosopic	8	8.16	14	14.00
Mesoprosopic	22	22.45	28	28.00
Leptoprosopic	46	46.94	34	34.00
Hyperleptoprosopic	22	22.45	22	22.00

The high frequency of mesorrhine is characteristic in both the sexes. The males do not present any broad and flat nose, while the females show about 6% of platyrrhine nose.

Both sexes are characterised by long faces. The next in order among the males is the mesoprosopic and hyperleptoprosopic face, while in the females this order is observed with only mesoprosopic faces. There are no available anthropometric measurements on this population group from Manipur or from Assam.

Manipuri Nagas

III. KABUI NAGA

The mean values of stature, cephalic index, nasal index and facial index of the Kabui Nagas are given below.

TABLE 3.10

Mean Values of Four Important Physical Traits in Kabui Nagas

Sex	*No.*	*Stature*	*C.I.*	*N.I.*	*F.I.*
Males	155	1571.58±3.72	74.98±0.18	75.31±0.49	89.01±0.31
Females	100	1480.61±6.86	76.78±0.48	72.90±0.96	86.00±0.68

The frequency distribution of these four physical traits are classified after Martin and Saller (1959).

Both sexes are found with short stature. The females possess a few cases of above medium to tall which is not found with the males.

Both sexes are identified with dolichomesocephalic type of heads. The females possess a few instances of broad and flat heads.

Both sexes have mesorrhine noses. The females possess about 6% broad and flat noses.

TABLE 3.11

Classification of Stature

Classification	*Males* *f*	*Males* %	*Females* *f*	*Females* %
Very short	5	3.23	2	2.04
Short	100	64.52	58	59.18
Below medium	40	25.81	16	16.33
Medium	10	6.45	18	18.37
Above medium	—	—	2	2.04
Tall	—	—	2	2.04

TABLE 3.12

Classification of Cephalic Index

Classification	*Males* *f*	*Males* %	*Females* *f*	*Females* %
Hyperdolichocephalic	10	6.45	6	6.00
Dolichocephalic	90	58.06	46	46.00
Mesocephalic	55	55.48	44	44.00
Brachycephalic	—	—	2	2.00
Hyperbrachycephalic	—	—	2	2.00

TABLE 3.13

Classification of Nasal Index

Classification	*Males*		*Females*	
	f	%	*f*	%
Leptorrhine	25	16.13	30	30.00
Mesorrhine	130	83.87	64	64.00
Chamaerrhine	—	—	6	6.00

TABLE 3.14

Classification of Facial Index

Classification	*Males*		*Females*	
	f	%	*f*	%
Hypereuryoprosopic	—	—	2	2.00
Euryoprosopic	25	16.13	14	14.00
Mesoprosopic	30	19.35	28	28.00
Leptoprosopic	70	45.16	34	34.00
Hyperleptoprosopic	30	19.35	22	22.00

Both sexes are having leptoprosopic faces. The next in order is mesoprosopic in the females while in males both mesoprosopic and hyperleptoprosopic faces occur in equal frequencies.

There exists two references on this population group. Waddel (1900) recorded average stature (2 subjects) as 1674.00 mm while Raychaudhuri and Chakladar (1957) measured 100 subjects and recorded the following mean values:

Mean stature=1558.7 mm
Mean cephalic index=77.48
Mean nasal index=67.95
Mean facial index=86.42

Our present measurements when compared with those of Chakladar and Raychaudhuri (1957) almost agree with each other. It appears that there was an increment of stature during these years by 13 mm.

IV. TANGKHUL NAGAS

The mean values of four important morphometric characters of the two sexes of the Tangkhul Nagas are mentioned below:

TABLE 3.15

Mean Values of Four Characters in Two Sexes

Sex	*No.*	*Stature*	*C.I.*	*N.I.*	*F.I.*
Males	106	1604.08±7.57	76.62±0.37	75.18±0.81	87.32±0.52
Females	108	1603.49±6.85	76.27±0.47	70.63±0.91	89.21±0.77

The distribution and classification of these four traits according to Martin and Saller (1959) are tabled on the next page.

The males are short while females are below medium in height. Four individuals were found with very short stature among the males. The medium category of stature is found with 46% of the population while in females this value is more (about 63%). Both sexes have two subjects each with tall stature.

The males are mesocephalic while the females are dolichocephalic. Among the females about 8% of the heads show hyper-dolicho forms.

The males are found with mesorrhine while half of the females are classified as leptorrhine and the remaining half are found with 46% mesorrhine and 4% platyrrhine noses.

TABLE 3.16

Classification of Stature

Classification	*Males* *f*	%	*Females* *f*	%
Very short	4	3.85	—	—
Short	52	50.0	38	36.54
Below medium	20	19.23	40	38.46
Medium	14	13.46	16	15.38
Above medium	12	11.54	8	7.69
Tall	2	1.92	2	1.92

TABLE 3.17

Classification of Cephalic Index

Classification	*Males* *f*	%	*Females* *f*	%
Hyperdolichocephalic	—	—	8	7.41
Dolichocephalic	48	45.28	60	55.56
Mesocephalic	52	49.06	36	33.33
Brachycephalic	2	1.89	4	3.71

The males have in general mesoprosopic face with a tendency towards leptoprosopic face. The females, on the other hand, are characterised by hyperleptoprosopic face with a tendency towards leptoprosopic face. So 15-38% of females have only mesoprosopic face against 47.17% of males.

TABLE 3.18

Classification of Nasal Index

Classification	*Males* f	%	*Females* f	%
Leptorrhine	20	17.87	54	50.00
Mesorrhine	82	77.56	50	46.30
Chamaerrhine	4	3.77	4	3.70

TABLE 3.19

Classification of Facial Index

Classification	*Males* f	%	*Females* f	%
Hypereuryoprosopic	4	3.77	2	1.92
Euryoprosopic	10	9.43	4	3.85
Mesoprosopic	50	47.17	16	15.38
Leptoprosopic	36	33.96	34	32.69
Hyperleptoprosopic	6	5.66	48	46.15

Waddel (1900) measured 10 male Tangkhul Nagas, the values of which were not available in the literature.

Before we take up the general discussion on the anthropometric characters among the Manipuris and their hill associates, we are tempted to refer to the writings of Haddon (1929) and Guha (1951).

HADDON (1929)

"An analysis of the anthropological data of the Assam tribes seems to indicate that there are several constituent races which do not coincide with political groups and are lost sight of when one deals with the averages. It may be tentatively suggested that there is an ancient dolichocephalic platyrrhine type (pre-Dravidian), which is strong among the Khasi, Kuki Manipur, Miki Kachari etc., but is weaker among the Naga tribes. There is reason to believe that a Nesiot element (dolichocephalic mesorrhine) is strong among the Naga and other hill tribes." (p. 116)

GUHA (1951)

"South of the Naga Hills is the state of Manipur and towards the West, the 'excluded' area of North Cachar, along the fringe of which the Naga settlements continue."

"From Manipur the tribal territory extends through the Lushai Hills, the hilly parts of Tipperah and the Chittagong Hill tracts into the Arakan Yoma ranges."

"... the entire tract forms a single geographical and ethnical unit closely knit in race and culture." (p. 22).

"Indications are not wanting also that even in the sub-Himalayan and the North-Eastern frontier countries where the Tibeto-Chinese speaking tribes are encountered at present there is an underlying sub-stratum of this element (Proto-Australoid) in several places."

"The second Mongoloid element is of medium stature with a skin-colour darker than that of the Tibetan but lighter than of the Palae-Mangoloid."

"Its chief distinguishing feature, is the form and shape of the head. The back of the head is neither flattened nor rounded as among the truly brachycephalic races, but bulges outwards. The breadth is also smaller and its proportion to the length is less. Though the actual values of the cephalic index often come within mesocephaly the essential long-headed character of the race cannot be doubted" (p. 33).

It will be seen from the table on the next page that the Tangkhul Nagas living in the eastern hills of Manipur are characterised by below-medium stature while the Kabuis are found with short stature. Probably this is due to their living in the hills. So far the cephalic indices are concerned the Tangkhuls are found with medium form of head while the Kabuis are found with long and

TABLE 3.20

Comparison of Four Important Physical Traits between the Manipur Nagas and the Nagaland Nagas

Population	*Stature*	*C.I.*	*N.I.*	*F.I.*
I. Manipuri Nagas: (All Males)				
a) Tangkhul Naga (106)	1604.08 mm	76.62	75.18	87.32
b) Kabui Naga (155)	1571.58 mm	74.89	75.31	89.01
II. Nagaland Nagas: (All Males)				
a) Angami (85)	1650.8 mm	76.65	74.59	89.01
b) Ao (77)	1580.6 mm	80.26	81.31	83.84
c) Kabui (100)	1558.7 mm	77.48	67.95	81.76
d) Konyak (94)	1583.2 mm	77.39	85.66	86.42
e) Lhota (41)	1620.0 mm	79.04	88.48	78.65
f) Sema (100)	1603.8 mm	79.31	71.18	86.33
g) Rengma (19)	—	75.69	80.79	—

narrow forms of head. Probably this is genetic in nature. Both are characterised by mesorrhine. The Tangkhuls are found with mesoprosopic faces while the Kabuis possess leptoprosopic faces. This is certainly due to their different ethnic origin and genetic component. Out of the seven Naga groups available for comparison, the Angamis are characterised by medium stature; the Aos, the Kabuis and the Konyaks are found with short stature. The present group of Kabui Nagas from Manipur could easily be grouped with them. The Lhotas, and the Sema Nagas are grouped conveniently with those of the Tangkhuls from Manipur.

There exists no broad-headed dominated Naga group either in

Manipur or in Nagaland. The Gengmas and the Kabuis of Manipur are found with long and narrow heads while the remaining six groups from Nagaland and the Tangkhuls from Manipur are grouped together having medium shaped heads. Both the Manipuri Nagas have mesorrhine noses and in this case they could easily be grouped with those of the Angamis, Aos, Semas and the Rengmas. The Kabui Nagas of Nagaland are having leptorrhine, while the Konyak as well as the Lhota Nagas both from Nagaland are found with broad and flat noses.

The Lhota Nagas are identified with leptoprosopic faces; the Kabuis and the Semas alongwith the Tangkhuls are having in general mesoprosopic faces; the Angamis and the Aos both are found with broad faces while the Konyak Nagas have very few broad faces. The Kabui Nagas of Manipur differ completely from all these groups having hyperleptoprosopic faces which are lacking with any other population group.

The Kabui Nagas of Manipur completely agree with Haddon's (1929) classification, while there exists some exception in the head form of the Tangkhul Nagas although the nasal index remains unchanged. Similarly, most of the Naga groups fall within the classification of Haddon most conveniently.

Risley's (1903) classification on the Nagas as well as the peoples in the adjoining areas do not clearly agree with the values.

Many years have elapsed since these scholars undertook the anthropometric investigations on these populations. The changes in the present morphometric features may suitably be attributed to a number of factors, both genetic and non-genetic.

APPENDIX

ANTHROPOMETRIC MEASUREMENTS OF THE PEOPLES OF MANIPUR

I.	*Kabui Naga* (*M*) :		102
	Waddel, L.A.	=	2
	Chakladar, H.C.	=	100
II.	*Koms* (*M*) :		50
	Das Gupta and Basu	=	50
III.	*Kuki Purum* (*M*):		60
	Das, T.C.	=	60
IV.	*Kuki Thadou* (*M*) :		60
	Hutton, J.H.	=	60
V.	*Kuki Thadou* (*F*) :		15
	William Shaw	=	15
VI.	*Tangkhul Naga* (*M*) :		10
	Waddel, L.A.	=	10
VII.	*Meitei* (*M*) :		15
	Waddel, L.A.	=	15

TRIBES OF NAGALAND

1.	*Angami Naga* (*M*) :		185
	Waddel, L.A.	=	16
	Hutton, J.H.	=	59
	Dixon, R.B.	=	10
	Singh, Sarajit	=	100
2.	*Ao Naga* (*Hatigoria*) (*M*):		51
	Waddel, L.A.	=	17
	Hutton, J.H.	=	34
3.	*Ao Naga* (*Chongli*) (*M*) :		71
	Hutton, J.H.	=	71
4.	*Ao Naga* (*Mongsen*) (*M*) :		23
	Hutton, J.H.	=	23
5.	*Asringia Naga* (*M*) :		3
	Waddel, L.A.	=	3
6.	*Chang Naga* (*M*) :		13
	Hutton, J.H.	=	13

7. *Kabui Naga* (*M*) :		100
Chakladar, H.C.	=	100
8. *Konyak Naga* (*M*) :		22
Hutton, J.H.	=	22
9. *Lhota Naga* (*M*) :		41
Waddel, L.A.	=	4
Hutton, J.H.	=	37
10. *Phom Naga* (*M*) :		1
Hutton, J.H.	=	1
11. *Rengma Naga* (*M*) :		19
Hutton, J.H.	=	19
12. *Sema Naga* (*M*) :		142
Hutton, J.H.	=	40
Dixon, R.B.	=	2
Singh, Sarajit	=	100
13. *Sema Naga* (*M*) :		8
Waddel, L.A.	=	8
14. *Sumdam Naga* (*M*) :		11
Waddel, L.A.	=	11
15. *Naga* (*M*) :		404
Hutton, J.H.	=	238
Kaufmann, H.E.	=	166
16. *Naga* (*F*) :		28
Kaufmann, H.E.	=	28

NAGA (238) Males :

Sengupta, S. 1951 — Measurements taken by Hutton, J.H. The Naga group consists of:

Ao Mongsen	= 23	
Ao Chongli	= 71	*Mean C.I.* : 78.7
Chang	= 16	*Mean N.I.* : 82.8
Sema Naga	= 41	
Angami Naga	= 28	
Konyak Naga	= 22	
Lhota Naga	= 37	

NAGA (166 Males) :

Olivier, G. 1956-58 — Measurements taken by Kaufmann, H.E.

Mean stature : 1590.09 mm ± 4.45

Mean C.I.	:	78.25	± 0.238
Mean N.I.	:	81.90	± 0.605
Mean F.I.	:	79.65	± 0.327

NAGA (28 Females):
Olivier, G. 1956-58
No further information.

ANGAMI NAGA (16 Males):
Waddel, L.A. 1900
Mean Stature : 1639 mm
Mean C.I. : 78.69
Mean N.I. : 82.54

ANGAMI NAGA (59 Males):
Hutton, J.H. 1921
Mean C.I. : 76.0
Mean N.I. : 79.0

ANGAMI NAGA (10 Males):
Hutton, J.H. 1921 Measurements taken by R.B. Dixon.
Mean Stature : 1648.50 mm
Mean C.I. : 77.53
Mean N.I. : 73.53
Mean F.I. : 84.67

ANGAMI NAGA (100 Males):
Singh, S. 1935
Mean Stature : 1654.3 mm ± 3.30
Mean C.I. : 76.67 ± 0.30
Mean N.I. : 68.93 ± 0.43

ANGAMI NAGA (28 Males):
Sengupta, S. 1951
Mean C.I. : 76.0
Mean N.I. : 80.6 Measurements taken by J.H. Hutton.

ANGAMI NAGA (85 Males):
Ray Chaudhuri, T.C. (Unpublished) : Based on the measurements of Dixon, Waddel and Hutton.

Mean Stature : 1650.80 mm
Mean C.I. : 76.65
Mean N.I. : 74.59
Mean F.I. : 83.84

AO NAGA (17 Males):
Waddel, L.A. 1900
Mean stature : 1566 mm
Mean C.I. : 80.11
Mean N.I. : 84.52

AO NAGA (34 Males):
Hutton, J.H. 1921
Mean C.I. : 81.0
Mean N.I. : 80.0

AO NAGA (CHONGLI) (71 Males):
Sengupta, S. 1951
Mean C.I. : 79.2 Measurements taken by J.H. Hutton.
Mean N.I. : 81.1

AO NAGA (MONGSEN) (23 Males):
Sengupta, S. 1951 Measurements taken by J.H. Hutton.
Mean C.I. : 78.8
Mean N.I. : 80.7

AO NAGA (77 Males):
Ray Chaudhuri, T.C. (Unpublished)
Mean stature : 1580.60 mm
Mean C.I. : 80.26
Mean N.I. : 81.31
Mean F.I. : 81.76

ASRINGIA NAGA (3 Males):
Waddel, L.A. 1900
Stature : 15557 mm.

ASRINGIA NAGA (3 Males):
Ray Chaudhuri, T.C. (Unpublished)
Mean stature : 1557 mm
Mean C.I. : 77.01
Mean N.I. : 84.94
Mean F.I. : 82.02

NAGA CHANG (13 Males):
Hutton, J.H. 1921
Mean C.I. : 79.0
Mean N.I. : 80.0

NAGA CHANG (16 Males):
Sengupta, S. 1951 Measurements taken by J.H. Hutton.
Mean C.I. : 77.5
Mean N.I. : 79.1

NAGA CHANG (29 Males):
Ray Chaudhuri, T.C. (Unpublished)
Mean C.I. : 77.2
Mean N.I. : 79.5

NAGA KABUI (2 Males):
Waddel, L.A. 1900
Mean stature : 1674

NAGA KABUI (100 Males):
Chakladar, H.C. and Ray Chaudhuri, T.C. 1936 : 1957
Mean stature : 1558.70 mm ± 3.72
Mean C.I. : 77.48 ± 0.18
Mean N.I. : 67.95 ± 0.49 Measurements taken
Mean F.I. : 86.42 ± 0.31 H.C. Chakladar.

NAGA KONYAK (22 Males):
Hutton, J.H. 1921
Mean C.I. : 70.0
Mean N.I. : 89.0

NAGA KONYAK (94 Males):
Sengupta, S. 1951
Mean stature : 1580.2 mm
Mean C.I. : 77.6
Mean N.I. : 89.4
Measurements taken by J.H. Hutton.

NAGA KONYAK (94 Males):
Ray Chaudhuri, T.C. (Unpublished)
Mean stature : 1583.2 mm
Mean C.I. : 77.39
Mean N.I. : 85.66
Mean F.I. : 78.65

NAGA LHOTA (37 Males):
Hutton, J.H. 1921
Mean C.I. : 79.0
Mean N.I. : 88.0

NAGA LHOTA (4 Males):
Waddel, L.A. 1900
Mean Stature : 1620.0 mm
Mean C.I. : 77.0
Mean N.I. : 79.0

NAGA LHOTA (37 Males):
Sengupta, S. 1951
Mean C.I. : 79.3
Mean N.I. : 90.0

NAGA LHOTA (41 Males):
Ray Chaudhuri, T.C. (Unpublished)
Mean Stature : 1620 mm
Mean C.I. : 79.04
Mean N.I. : 88.48
Mean F.I. : 97.37

NAGA PHOM (1 Male):
Hutton, J.H. 1921
Mean C.I. : 81.87
Mean N.I. : 75.55

NAGA RENGMA (19 Males):
Hutton, J.H. 1921
Mean C.I. : 79.0
Mean N.I. : 82.0

NAGA RENGMA (8 Males):
Waddel, L.A. 1900
Mean Stature : 1501 mm

NAGA RENGMA (19 Males):
Ray Chaudhuri, T.C. (Unpublished)
Mean C.I. : 75.69
Mean N.I. : 80.79

SEMA NAGA (40 Males):
Hutton, J.H. 1921
Mean C.I. : 79.0
Mean N.I. : 80.0

SEMA NAGA (8 Males):
Waddel, L.A. 1900
Mean Stature : 1501 mm

SEMA NAGA (2 Males):
Hutton, J.H. 1921
Mean stature : 1657.50 mm
Mean C.I. : 78.69
Mean N.I. : 74.40
Mean F.I. : 86.33

SEMA NAGA (100 Males):
Singh, S. 1935
Mean Stature : 1603.80 mm ± 4.00
Mean C.I. : 79.31 ± 0.22
Mean N.I. : 71.18 ± 0.50

SEMA NAGA (40 Males):
Ray Chaudhuri, T.C. (Unpublished)

C.I.	:	Hyperdolicho-Dolicho—	7	17.50
		Meso	18	45.00
		Brachy	14	35.00
		Hyperbrachy	1	2.50
N.I.	:	Leptorrhine	4	10.00
		Mesorrhine	26	65.00
		Chameorrhine	9	22.50
		Hyper	1	2.50

SEMA NAGA (41 Males):
Sengupta, S. 1951
Mean C.I. : 79.9
Mean N.I. : 79.5

NAGA SUMDAM (11 Males):
Waddel, L.A. 1900
Stature : 1580 mm

NAGA SUMDAM (11 Males):
Ṛay Chaudhuri, T.C. (Unpublished)
Stature : 1850 mm
C.I. : 76.09
N.I. : 92.86
F.I. : 75.00

MEITEI MANIPURI (15 Males):
Waddel, L.A. 1900
Mean stature : 1636mm

KOM (50 Males):
No Informations

KUKI PURUM (60 Males):
Das, T.C. 1945

Stature	:	1549.00mm	± 4.2
C.I.	:	77.24	± 0.28
N.I.	:	70.84	± 0.58

KUKI THADOU (60 Males):
Shaw, William. 1927-28

Stature	:	1608.7 mm
C.I.	:	76.50
N.I.	:	80.70

KUKI THADOU (60 Males):

Stature	1603.00 mm	± 4.7
C.I.	76.64	± 0.30
N.I.	0.95	± 0.07

KUKI THADOU (15 Females):
Shaw, William 1927-28

Stature	:	1466.6 mm
C.I.	:	76.60
N.I.	:	81.10

References

1. Chakladar, H.C. and T.C. Ray Chaudhuri (1957): *The Kabui*: *Man in India*, 37: 192-198.
2. Datta, B.N. (1939): "Anthropological notes on some Assam castes and tribes", *Anth. Papers*, University of Calcutta, New series No. 5: pp. 1-26.
3. Das, T.C. (1945) : *The Purums* : *An old Kuki tribe of Manipur*, University of Calcutta, 25-39.
4. Guha, B.S. (1935): "The racial affinities of the peoples of India", in *Census of India*, 1931, vol. I., India Pt. IIIA, pp. 2-22.
5. Guha, B.S. (1951) "The Indian aborigines and their administration", *J.A.S.B.* XVII : 1 : 19-44.
6. Haddon, A.C. (1929): *The races of man and their distribution*, Cambridge University Press, London.
7. Hutton, J.H. (1921): *The Angami Nagas*, London. Appendix XI : pp. 423-38.
8. Martin, R. and K. Saller (1959): *Lehrbuch der Anthropologie*, Vols. I and II, Gustav Fischer Verlag, Stuttgart.
9. Olivier, G. (1956): "Anthropology of the Naga head hunters of Assam", *Bull. Soc. Anthrop*, 6: 6: 373-417.
10. Olivier, G. (1958): "Physical Anthropology of the Nagas of Assam", *Man in India*, 38: 105-110.
11. Ray Chaudhuri, T.C., Unpublished cf. P. Gupta and P.C. Datta: *Anthropometry in India*, Calcutta, 1966.

12. Risley, H.H. (1891): "The tribes and castes of Bengal", *Anthropometric Measurements.* Vols. I & II, Calcutta.
13. Risley, H.H., (1903): "Seriations of the Mongoloid-Dravidian type Koch", in *Census of India*, 1901; Vol. I., India, Pt. I, *Report of ethnographic appendices*, pp. 28.
14. Shaw, William. (1927-28): "Notes on the Thadou Kukis", *Proc. Royal. Soc., Bengal*, N.S. Vol. 24: Appendix H: pp. 159-162.
15. Sengupta, S. (1951): "A note on Naga Anthropometry", *J.A.S.B.* 17 : Science 1-5.
16. Singh, S., (1935) : "A contribution to Angami and Sema Somatology", *J.A.S.B.* Vol. Science, 93-97.
17. Waddel, L.A. (1900) : "The tribes of Brahmaputra valley: a contribution on their physical types and affinities", *J.A.S.B.*, 69 : III : 1-127, Calcutta.

4

Dermatoglyphics Among Manipur Population Groups

Dermatoglyphic research in North East India is more than two decades old by now. Dermatoglyphic characters of the various populations have been reported by different scholars from time to time. Bhattacharjee (1955) studied the Abors; Kumar examined (1954) the Galongs; Das and Bhagabati (1962 : 1964) studied the Khamyang; Srivastava (1969) the Miris. Das (1974) has presented an excellent review on the Dermatoglyphic research in North East India.

This chapter examines afresh the finger and palm prints of the Manipur tribes and castes. A study examining the relationship between Dermatoglyphics and the Ulcers of two different types has been included as an appendix. This article was translated by Prof. M. C. Chamla and was published in L'Anthropologie, vol : 84; No : 1 (1980).

The North-Eastern region of India provided shelter to numerous Mongoloid tribes. These tribes have been conveniently grouped under the various divisions on the basis of their cultural similarity, linguistic affinity and common territory.

Chakravartti (1963) undertook a Dermatoglyphic survey on the Manipur population groups.

A. FINGER PRINTS

The Mongolian samples (Chakravartti, 1963) show nearly equal Whorl-Loop ratio (50 W : 50 L) and track along the

Mongolian Map of India, about which there is less diversity of opinion. The interesting point is that some of them have 55 : 45 (W : L) probably verging towards Australoids of South East and some have 45 : 55 (W : L) in the Eastern frontier and they may have tendency towards some North Indian people who show considerable lowering of Whorl frequency (about 30—35%).

The North-Eastern people throw some light on the Mongolian question in India. A high percentage of Whorl and longitudinal main lines known to be of Australoid character are very common in the Eastern border among the Naga and some Kuki-Chin tribe (Chakravartti and Mukherjee, 1961), which is similar to Eickstedt's (1934) primitive palae-Mongoloid type. No difference between the long headed and broad headed Mongoloid of Guha (1935) could be traced so far. Western tribes of North-Eastern region are of progressive type as far as Dermatoglyphics is concerned. The high caste elements of Western ethnic origin could be traced as far as Lakhimpur and Sibsagar districts of Assam. The disputed difference of the Meiteis and the Bishnupriyas—the two sets of Manipuri communities—could be clarified by the fact that infiltration of Bengali high castes within the latter is interesting. As a whole the Bishnupriya Manipuris do not seem to have any Mongoloid strain. The Mongoloid populations of India thus present three sets of Whorl-Loop ratio:

(a) Nearly equal Whorl-Loop ratio (50 : 50)
(b) Peoples verging towards Australoids of South East (55 : 45; W : L)
(c) Tendency towards North Indian people, who show considerable lowering of Whorl frequency (45 : 55; W : L).

It will be seen from Table 4.1 that the Bishnupriya Manipuris are characterized by high percentage of arches in both sexes. Both the Manipuris are different from each other on the basis of finger print patterns. As stated earlier that the Bishnupriyas have Bengali high caste infiltration because of their present and earlier close association with the Bengalis of Silchar district of Assam.

The Meitei males present W>L while the females have just the reverse L>W. Both sexes are characterized by high percentage

TABLE 4.1

Finger Prints of the Manipuri Population Groups

Population	*Sex/No.*	*Whorl*	*Loop Radial*	*Loop Ulnar*	*Arch*
1. Meitei	M—118	51.95	2.12	44.07	1.86
	F—128	43.36	1.95	33.20	1.48
2. Hmar	M—100	48.60	3.50	46 80	1.10
	F—148	41.62	2.03	55.95	0.41
3. Chothe	M— 84	49.05	0.48	50.00	0.48
	F— 68	38.82	2.35	58.92	—
4. Moyon Monsong	M—116	48.28	2.76	48.62	0.34
	F— 62	40.00	2.58	56.13	1.29
5. Gangte	M—100	46.00	2.00	50.00	2.00
	F—100	45.00	1.50	52.50	1.00
6. Bishnupriya Manipuri	M—110	44.09	1.36	49.91	4.64
	F—110	45.55	0.54	49.64	4.27
7. Kabui	M—164	44.29	3.54	52.07	0.12
	F—118	36.44	3.73	58.98	0.85
8. Anal	M— 45	34.44	3.33	60.00	2.22
	F— 30	45.00	5.00	50.00	—
9. Zeu	M— 84	38.81	3.33	57.86	—
	F— 30	58.33	6.67	35.00	—
10. Veipei Kuki	M—148	45.54	2.16	50.68	1.62
	F— 86	46.28	1.40	50.93	1.40
11. Thadou Kuki	M—100	43.80	1.60	53.60	1.00
	F—124	40.48	0.97	55.65	2.90
12. Lushai	M—134	41.64	3.73	53.28	1.34
	F—124	41.13	3.06	52.58	3.23
13. Mao Naga	M—153	55.10	1.96	42.81	0.13
	F—102	51.67	2.06	45.98	0.29
14. Tangkhul	M—270	53.45	2.15	43.00	1.40
	F—172	55.70	1.16	41.63	1.51
15. Paite	M—101	51.39	2.67	44.26	1.68
	F—150	47.20	2.00	50.40	0.40

of Radial Loops. The Bishnupriyas, on the other hand, are having L>W in both sexes. In this respect, the Meitei females have close similarity with those of the Kabui Nagas of the valley who also have L>W in their digital distribution among the two sexes. The Meitei males, on the other hand could be grouped alongwith the Tangkhul and Mao Nagas (both sexes) who have W>L. The distribution of Radial Loops as well as the arches also presents uniformity among the three Manipuri Naga groups. The Mao Nagas living in the border of Nagaland and Manipur and inhabiting the hilly area are found in one extreme, while the other Manipuri Naga group, the Tangkhuls, inhabiting the Eastern Hill areas occupy the other extreme, in so far as the distribution of the arches and the Radial Loops are concerned. The Maos have about 2% Radial Loop and the Tangkhuls have 1-2% of Radial Loops while the Kabuis occupying the intermediate position between these three communities and living in the valley have about 3.5—4% of Radial Loops. Similar trend is also noticed in the case of arches.

The Maos have very low occurrence of arches (about 0.1—0.2%) in one extreme, while the Tangkhuls have about 1.5% on an average. The Kabuis occupying the intermediate position have 0.1—0.8% of arches.

The digital patterns of the two Manipuri groups and the three Manipuri Naga groups pose a general question. In all the five groups, the Whorl-Loop occurrence presents uniformity in both sexes excepting the Meitei females. They do not follow the male pattern like other groups. Does this mean that there was some effect of miscegenation among the Meitei females. The similar trend of the Meitei females alongwith the Kabuis as well as that of the Bishnupriya group is very interesting and needs further probe.

There are as many as nine Kuki groups and one Lusai group in the collected list of Table 4.1.

As will be seen, all the Kuki groups as well as the Lusai group present l>W as comparative occurrences of these two digital patterns. The only deviations are noted in the females of Zou and the males of Paite. It was earlier suggested by Chakravartti (1963) that the general trend of the Whorls is to rise gradually towards the South of the Indian subcontinent (with a trend towards Eastern part) and loops in the reverse direction.

It is, therefore, expected that in the Manipur population groups we will come across higher occurrence of loops compared to Whorls. The occurrence of arches in the Kuki groups of Manipur is very low. The two major Kuki groups—the Thadou and the Veipei—are characterized by 2% and 1.5% of arches in the two sexes respectively. The Lusais have a little higher frequency of arches. All other sub-groups of Kukis have less than 1% of arches. The Zou group does not have any arches at all in the two sexes.

The Radial Loop occurs in about 3% among the Lusai groups. The Thadous and the Veipei Kukis have approximately half of its occurrence. The high frequency of Radial Loops among the Zou, Anal may be due to their low occurrence of arches. Besides, the other sub-groups have a range of occurrence of Radial Loops from 1.5 to 2.7%.

Tables 4.2–4.7 describe the Mongoloids of North East India with three different sets of Whorl-Loop ratio.

Chakravartti (1963), on the basis of the relative ratio of Whorls and Loops, classified three groups in India.

1.	W 60 : L 40	Australoid
2.	W 40 : L 60	Mundari-Speakers
3.	W 50 : L 50	Mongoloids

The values in Table 4.7 again confirms the classification by Chakravartti (1963). The Mongoloids and the Mundaris appear to stand close to one another in respect of papillary patterns than either of them to the Australoids. This evidence was earlier confirmed by Anthropometry. So, it is not unlikely that a few Mongoloid population groups of North Eastern India have shown certain Mundari-speaking patterns in respect of digital prints.

Most probably this may be due to their extreme miscegenation. To sum up, so far the digital patterns are concerned among the population groups of Manipur, we have seen that all the three sets of Whorl-Loop ratios are present alongwith a new set having close similarity with the formula of the Mundari-speakers.

TABLE 4.2

Nearly Equal Whorl Loop Ratio (50 : 50 : W : L) (*Males*)

Sl. No. Populations	*No.*	*Whorl*	*Loop R*	*Loop U*	*Arch*
1. Hill Garo	76	50.76	0.93	46.55	1.74
2. Plain Garo	94	47.96	1.39	47.57	2.98
3. Rabha	295	50.66	1.55	46.24	1.55
4. Galong	152	46.00	2.07	49.92	2.00
5. Miril	100	49.29	1.61	48.00	0.90
6. Meitei Manipuri	118	51.95	2.12	44.07	1.86
7. Hmar	100	48.60	3.50	46.80	1.10
8. Chothe	84	49.05	0.48	50.00	0.48
9. Moyon Monsong	116	48.28	2.76	48.62	0.34
10. Gangte	100	46.00	2.00	50.00	2.00
11. Lalung	106	51.04	1.70	45.75	1.51
Total	1441	49.00	1.83	47.59	1.58

TABLE 4.3

Nearly Equal Whorl-Loop Ratio (*Females*)

Sl. No. Populations	*No.*	*Whorls*	*Loop R*	*Loop U*	*Arch*
1. Mao Naga	102	51.67	2.06	45.98	2.09
2. Tangkhul Naga	172	55.70	1.16	41.63	1.51
3. Paite	150	47.20	2.00	50.40	0.40
4. Bodo	104	47.88	1.54	47.69	2.89
5. Abor	163	45.16	1.49	50.87	2.48
Total	691	49.52	1.65	47.32	1.51

TABLE 4.4

Tendency towards North Indian people, who show considerable lowering of Whorl frequency. (45 : 55 W : L) (*Males*)

Sl. No.	*Populations*	*No.*	*Whorl*	*Loop R*	*Loop U*	*Arch*
1.	Ahom	112	43.48	2.59	52.50	1.43
2.	Airi	110	40.28	1.36	57.27	1.09
3.	Bishnupriya Manipuri	110	44.09	1.36	49.91	4.64
4.	Mech	72	40.28	1.94	55.28	2.50
5.	Hajong	75	44.68	1.73	51.96	1.63
6.	Rajbanshi	130	43.11	2.02	51.12	3.73
7.	Kabui Naga	164	44.27	3.54	52.07	0.12
8.	Veipei Kuki	148	45.54	2.16	50.68	1.62
9.	Thadou Kuki	100	43.80	1.60	53.60	1.00
10.	Lushai	134	41.64	3.73	53.28	1.34
11.	Khasi	260	41.73	2.55	53.31	1.38
	Total	1657	44.00	2.25	52.90	1.85

B. PALM PRINTS

In order to display the degree of transversality in their coursing, the three important palmar main line formulae (Wilder, 1922) only have been grouped and considered.

The Meitei Manipuris present different modal types for the three main line formulae D, C, B, in the two sexes.

The males show 7.5.5., 9.7.5., 11.9.7., while what the females have, was not found in any other population groups of Manipur. The three Manipuri Nagas (both sexes) follow the same modal

TABLE 4.5

43 : 55 (W : L) (*Females*)

Sl. No.	*Populations*	*No.*	*Whorl*	*Loop R*	*Loop U*	*Arch*
1.	Rabha	295	44.98	0.94	49.77	4.31
2.	Miri I	58	44.42	1.37	52.63	1.56
3.	Meitei Manipuri	128	43.36	1.95	53.20	1.48
4.	Hmar	148	41.62	2.03	55.95	0.41
5.	Moyona Monsong	62	40.00	2.58	56.13	1.29
6.	Gangte	100	45.00	1.50	52.50	1.00
7.	Miri II	112	40.09	0.62	56.17	3.12
8.	Bishnupriya Manipuri	110	45.55	0.54	49.64	4.27
9.	Anal	30	45.00	5.00	50.00	—
10.	Veipei Kuki	86	46.28	1.40	50.93	1.40
11.	Thadou Kuki	124	40.48	0.97	55.65	2.90
12.	Lushai	124	41.13	3.06	52.58	3.23
13.	Zou	30	58.33	6.67	35.00	—
	Total	1407	44.26	2.16	51.50	1.10

type like that of the Meitei males. Among the Kuki-Chin groups, some deviation is noticed. Both the Thadou and the Veipei Kukis follow the formula of the Manipuri Nagas alongwith the other smallar sub-groups. The Lushai males show the model type of D, C, and B as 9.7.5. 7.5,5. 11.9.5. The females, on

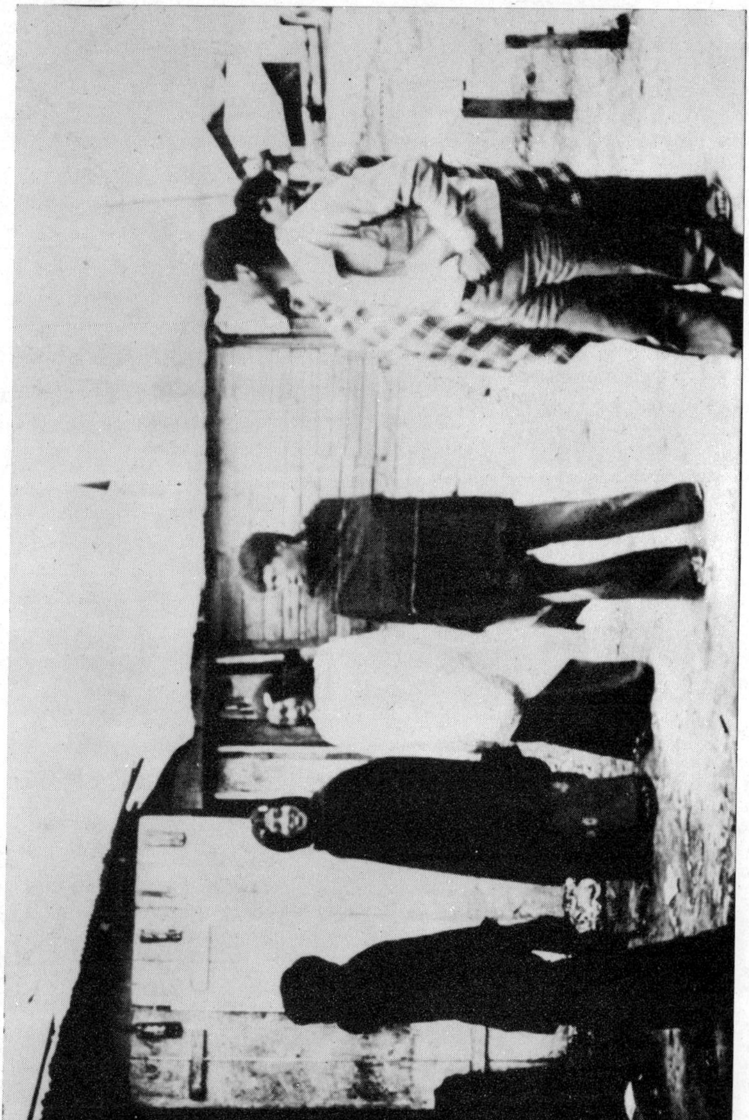

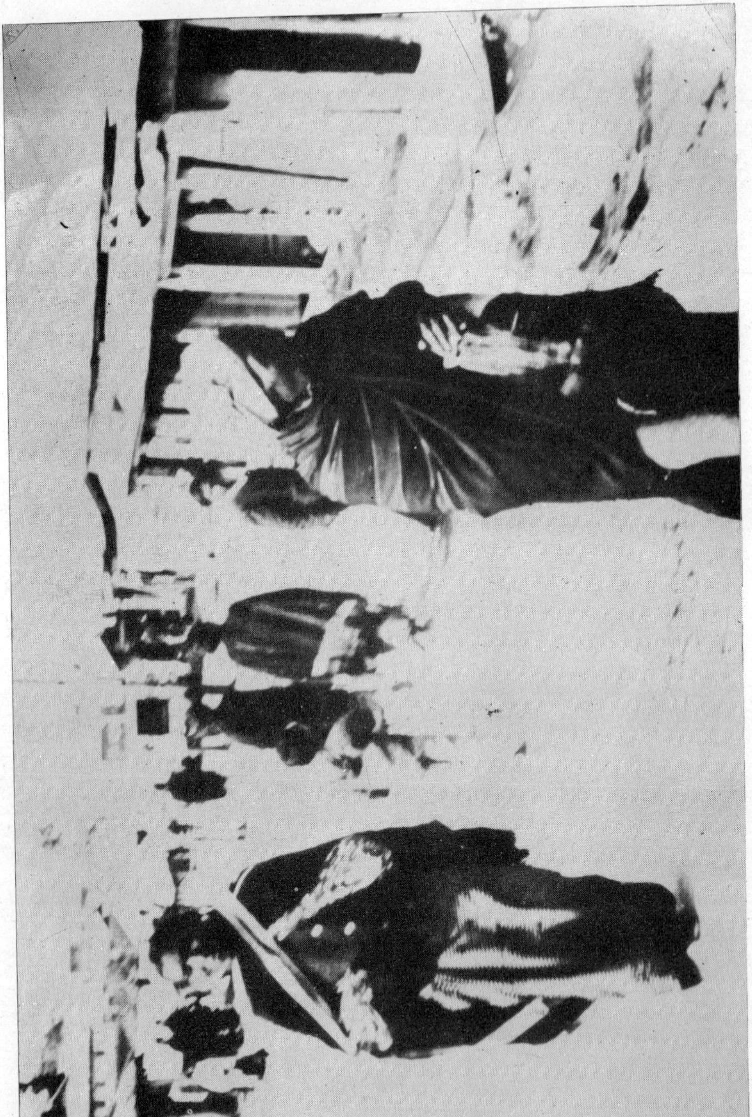

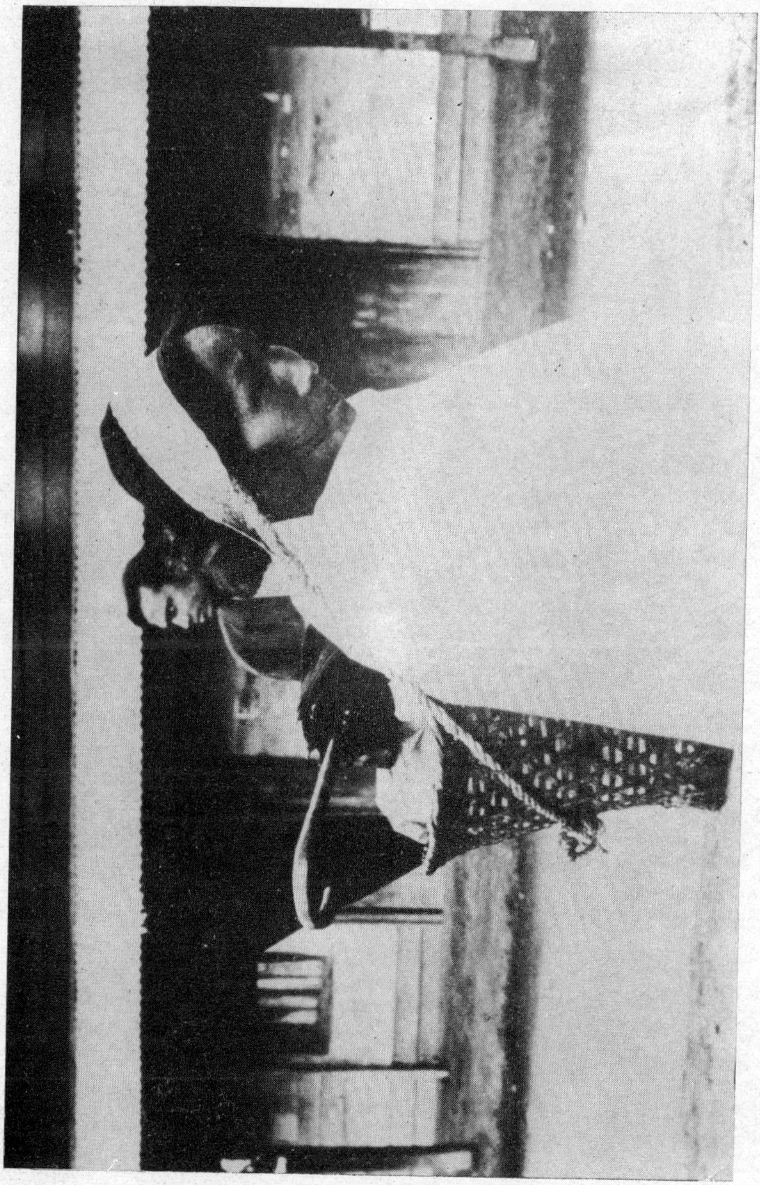

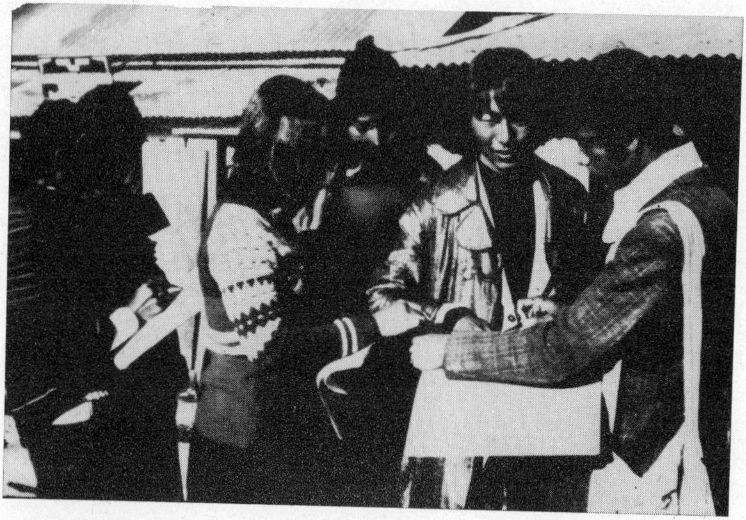

TABLE 4.6

People Verging Towards Australoids (55 : 45 W : L)
(*Males*)

Sl. No.	*Populations*	*No.*	*Whorl*	*Loop R*	*Loop U*	*Arch*
1.	Mao Naga	153	55.10	1.96	42.81	0.13
2.	Tangkhul	270	53.45	2.15	43.00	1.40
3.	Paite	101	51.39	2.67	44.26	1.68
4.	Kachari	109	54.66	0.55	42.86	1.84
5.	Bodo	138	55.22	1.16	42.46	1.16
6.	Abor	147	53.24	3.03	41.25	2.48
	Total	918	53.85	1.92	42.78	1.45

TABLE 4.7

Whorl Less Than 40%

Sl. No.	*Populations*		*No.*	*Whorl*	*Loop R*	*Loop U*	*Arch*
1.	Chothe	F	68	38.82	2.35	58.92	—
2.	Lalung	F	132	38.94	2.12	55.76	3.18
3.	Kalita	M	113	33.19	2.21	60.71	3.89
4.	Anal	M	45	34.44	3.33	60.00	2.22
5.	Zou	M	84	38.81	3.33	57.86	—
6.	Ahom	F	110	33.19	2.21	60.71	3.89
7.	Kalita	F	120	27.42	2.17	65.91	4.50
8.	Mech	F	50	38.80	2.40	56.80	2.00
9.	Kabui	F	118	36.44	3.73	58.98	0.85
10.	Khasi	F	256	35.41	2.83	58.18	3.90
	Total		1096	35.55	2.66	59.38	2.41

TABLE 4.8

Three Main Palmar Main Line Formulae among Manipur Groups

Sl. No.	*Populations*	*Sex*	*No.*	*11.9.7*	*9.7.5.*	*7.5.5.*
1.	Meitei	Male	118	11.44	32.63	40.25
		Female	128	20.31	19.53	48.83
2.	Hmar	Male	100	5.50	37.50	49.50
		Female	148	9.46	20.27	52.70
3.	Chothe	Male	84	19.05	28.57	42.86
		Female	68	11.76	20.59	64.71
4.	Moyon	Male	116	16.38	41.38	31.03
		Female	62	4.84	27.42	62.90
5.	Gangte	Male	100	12.00	29.00	60.00
		Female	100	7.50	25.00	65.00
6.	Kabui	Male	164	11.59	37.20	37.80
		Female	118	11.86	25.42	45.76
7.	Anal	Male	45	16.67	38.89	27.78
		Female	30	—	16.67	83.33
8.	Veipei	Male	148	12.16	27.03	53.38
		Female	86	3.49	23.26	62.79
9.	Thadou	Male	100	12.00	22.50	49.00
		Female	124	16.13	21.77	50.00
10.	Lusai	Male	134	15.67	38.06	29.85
		Female	124	10.48	31.45	49.19
11.	Zou	Male	84	3.57	16.67	59.92
		Female	30	8.33	8.33	75.00
12.	Mao Naga	Male	153	7.84	19.93	54.58
		Female	102	14.71	21.57	45.59
13.	Tangkhul	Male	270	14.50	29.75	45.25
		Female	172	13.37	26.75	44.19
14.	Paite	Male	101	9.90	34.65	39.11
		Female	150	8.67	21.33	46.67

the other hand, follow the Kukis in general. The Moyon males and the Anal males show close affiliation with those of the Lushai males so far as the main line formulae are concerned.

From this it can be deduced that all the Manipur population groups show the highest frequency of the 7.5.5. main line formulae followed by 9.7.5. and 11.9.7.

It may be stated that Chakravartti (1963) in his Dermatoglyphic map of India had shown that the Australoid males present 11.9.7 in highest frequency while the females show 7.5.5. main line in highest frequency. The predominance of formulae 7.5.5. among both the sexes of the Mundaris of two types and the Mongoloids of three concentration is noteworthy.

Tables 4.9—4.13 describe the percentile frequency of three main palmar main lines of the Mongoloids of North East India with special reference to Manipur.

The group having 50 : 50 (W : L) ratio in both their sexes are characterized by the following:

(a) 11.9.7=above 10%
(b) 7.5.5=more than 40%
(c) 9.7.5=within 30%

The group having 45 : 55 (W : L) in their males are characterized by:

(a) 11.9.7=above 10%
(b) 7.5.5=more than 40%
(c) 9.7.5=less than 30%

Their females have:

(a) 11.9.7=below 10%
(b) 7.5.5=more than 50%
(c) 9.7.5=within 20%

The group verging towards Australoid (55 : 45 W : L) is characterized by the following:

(a) 11.9.7=above 10%
(b) 7.5. =less that 50%
(c) 9.7.5=within 30%

TABLE 4.9

Nearly Equal Whorl-Loop Ratio (50 : 50)
(*Males*)

Sl. No.	*Populations*	*11.9.7*	*9.7.5*	*7.5.5*
1.	Rabha	4.20	21.36	28.30
2.	Meitei	11.40	32.63	40.25
3.	Hmar	5.50	37.50	49.50
4.	Chothe	19.05	28.57	42.86
5.	Moyon	16.38	41.31	31.03
6.	Gangte	12.00	29.00	60.00
	Total	11.43	31.74	41.74

TABLE 4.10

Nearly Equal Whorl-Loop Ratio (*Females*)

Sl. No.	*Populations*	*11.9.7*	*9.7.5*	*7.5.5*
1.	Mao Naga	14.71	21.57	45.49
2.	Tangkhul Naga	13.37	26.74	44.19
3.	Paite	8.67	21.33	46.67
	Total	12.25	23.21	45.48

TABLE 4.11

Tendency Towards 45 : 55 (W : L)
(*Males*)

Sl. No.	*Populations*	*11.9.7*	*9.7.5*	*7.5.5*
1.	Kabui	11.59	37.20	37.80
2.	Anal	16.67	38.89	27.78
3.	Zou	3.57	16.67	59.52
4.	Veipei Kuki	12.16	27.03	53.38
5.	Thadou Kuki	12.00	22.50	49.00
6.	Lushai	15.67	38.06	29.85
7.	Hajong	9.23	27.69	44.61
8.	Ahom	14.61	26.04	34.46
	Total	12.06	29.26	42.05

Rife (1963) has shown very well the ethnic variations in the palmar patterns. Compared with finger prints, palmar patterns show much lesser sex variations. The Mongoloids of North Eastern region vary widely in respect of hypothenar and the IV interdigital patterns from both the Australoids and the Mundaris.

TABLE 4.12

Tendency Towards 45 : 55 (W : L)
(*Females*)

Sl. No.	*Populations*	*11.9.7*	*9.7.5*	*7.5.5*
1.	Rabha	3.90	18.13	32.06
2.	Meitei	20.31	19.53	48.83
3.	Hmar	9.46	20.27	52.70
4.	Chothe	11.76	20.59	64.71
5.	Moyon	4.84	27.42	62.90
6.	Gangte	7.50	25.00	65.00
7.	Kabui	11.86	25.42	45.76
8.	Anal	—	16.67	83.33
9.	Zou	8.33	8.33	75.00
10.	Veipei	3.49	23.26	62.79
11.	Thadou	16.13	21.77	50.00
12.	Lushai	10.48	31.45	49.19
13.	Khasi	14.18	23.28	34.51
	Total	9.40	22.40	55.90

TABLE 4.13

Verging Towards Australoid (55 : 45 W : L)
(*Males*)

Sl. No.	*Populations*	*11.9.7*	*9.7.5*	*7.5.5*
1.	Mao Naga	7.84	19.93	54.58
2.	Tangkhul	14.50	29.75	45.25
3.	Paite	9.90	34.65	39.11
4.	Kachari	12.00	31.33	35.33
	Total	10.81	23.91	43.37

References

1. Bhattacharjee, P.N. (1955) : "Dermatoglyphic among the Abor," *Bull. Deptt. Anth. Govt. India* III : 1.
2. Chakravartty, M.R. (1963) : "Ethnic bearing of the Dermatoglyphic studies in India", *Acta Genet Med. Gemellol (Roma)* 12 : 4.
3. Das, B.M. and A.C. Bhagabati (1962) : "Investigations on the finger prints of the Khamyang", *J. Gauhati Univ*, XIII : 1.
4. Das, B.M. (1974) : *Contemporary Anthropology in North East India*, *Ed.* Dibrugarh University, Dibrugarh, Assam.
5. Guha, B.S. (1935) : "The racial affinities of the peoples of India", in *Census of India* Vol I, Pt. I, New Delhi.
6. Kumar, N. (1954) : "Dermatoglyphics of the Galong", *Bull. Anth. Deptt. Govt. India* 3 : 1.
7. Rife, D.C. (1963) : "Dermatoglyphics : Introductory remarks." *Proc. II International Congress Human Genetics*, Rome 1961. Vol. III 1927, Inst. G. Mendel Rome 1963.
8. Chakravartti, M.R. and D.P. Mukherjee (1961) : "Dermatoglyphics of the tribes and castes of Assam", *Science and Culture*, 27 : 2.
9. Srivastava, R.P. (1969) : "Finger Prints, Colour Blindness and Other Features in the Miris of Assam," *Man in India*, 49 : 398-403.

APPENDIX

Dermatoglyphics Studies in Duodenal and Peptic Ulcer

Translated from French:
Reprinted from L' Anthropologie vol : 84 : No. 1.
(1980) by Rama Chakravartti and M.R. Chakravartti.

INTRODUCTION

A good number of diseases have been studied to look for associations, with Dermatoglyphic parameters. The specific association, if confirmed, may serve as diagnostic aid. For example, a high atd angle and the presence of simian crease are already being used for the diagnosis of Down's syndrome. This study was undertaken in search for an association between Dermatoglyphic traits and the two types of ulcer. To the best of our knowledge Dermatoglyphics have not been studied in this type of cancer.

Many exogenous and endogenous factors are involved in the development of peptic ulcers. These factors are of individual importance for each subject and probably vary with time. Great importance is ascribed to hereditary factors. (Buckwalter, 1962).

McConnel (1966) has discussed the genetics of gastrointestinal disorders.

Study on the probable association between peptic ulcer (regardless of the type of ulcer) and ABO Bloodgroups reveals that a positive family history of peptic ulcer combined with blood group O is an important sign and significant characteristics of duodenal ulcer. Based on many well-documented publications from all over the world, a proven relationship is considered to exist between duodenal ulcer and blood group O.

Materials and Methods

The sample for the present study was drawn from the Regional Medical College Hospitals, Imphal, Manipur during the summer months of May-July, 1973. Finger and palm prints of 100 females suffering from the two types of ulcers were taken and compared with 100 normal females of comparable age group.

The patients as well as the controls were mostly from Imphal town and the neighbourhood. Both the samples were of mixed origin. The prints were recorded after Cummins and Midlo (1943).

Results and Discussion

Since no studies have been done in the past on this problem, we are unable to compare our data. The results of the present study are summerised in the following tables:

Finger Tip Patterns

TABLE 1.0

Percentile Frequency of Digital Patterns

Pattern type	*Patients (100)*	*Control (100)*
Arches	4.00	3.00
Loops Ulcer	56.00	60.00
Loops Radial	2.00	1.00
Whorls	38.00	36.00

X^2 1 = 0.29 not significant.

It will be seen from the above table that there was not much variation in the frequency of the digital patterns between the controls and the patients.

Palmar Patterns

TABLE 2.00

Frequency (%) of Presence of Patterns Hypothenar, Thenar/I, II, III & IV Interdigital Areas.

	Hypothenar	*Thenar/I*	*II*	*III*	*IV*
Patients	36.00	1.5	3.00	51.00	62.00
Controls	30.00	6.5	1.5	44.00	49.00

X^1 1=between Hypothenar and Thenar/I 3.62 n.s.
X^2 2=between II, III and IV Interdigital areas 0.33.
The differences in the above pattern areas are, however, not significant.

TABLE 3.0

Total Finger Ridge Count, Main Line Index, Pattern Intensity Index and Mean atd Angle.

Parameters	*Patients*	*Controls*
Total Finger Ridge Count	122.25	120.50
Main Line Index	8.16	7.96
Pattern Intensity Index	14.20	13.75
Mean atd Angle	42.00	44.00

There is a very slight increase in the total finger ridge count in the patients but the difference is not significant.

The differences between the mean of MLI, PII in the two series are found to be non-significant. The maximal atd angle was found to be non-significant.

Summary

The finger and palm prints of 120 females suffering from duodenal and peptic ulcers are analysed and the results are compared with those of 100 normal females. It appears that there are no differences between the two series.

References

1. Buckwalter, J.A. (1962): "Peptic Ulcer and the Bloodgroups", *Ann. New York Acad. Sci.* 99.
2. Cummins, H. and C. Midlo (1943) *Fingerprints, Palms and Soles*, Blackiston, Philadelphia.
3. McConnel, R.B. (1966) *The Genetics of Gastro-Intestinal Disorders*, O.U. Press, London.

5

Blood Groups in Manipur Population Groups

Pre-independent North-East India has only a very limited series of ABO blood group data. Mitra (1936) examined the ABO blood group of the Assamese of the plains alongwith the Angami and Lusai tribes from the hills. Basu (1938) and Mac-farlane (1941) screened two series of Khasi data in respect of ABO blood groups. British Association (1939) described two series of Naga blood groups.

Post-independent serological investigation in North East India was initiated by Das (1960). He reported the ABO blood group data of the Rabha tribe of Assam. Besides, he examined a number of tribes and caste populations of Assam and attempted an ethic affinity within these groups (1974). Compared to Assam and Nagaland, serological data on ABO or other blood groups of Manipur population groups are woefully inadequate. To our knowledge, there exists no other serological data on Manipur peoples till date.

ABO Blood Groups

In all, 1065 individuals (unrelated) belonging to four population groups of Manipur were screened for ABO blood groups.

It will be seen from the Table 5.0 that the blood group O predominates in the three, out of four population groups. The

TABLE 5.0

Distribution of ABO Blood Groups

Populations	*No.*	*O*	*A*	*B*	*AB*
Meiti (caste Hindu)	380	130	124	84	42
Muslim (muslim)	227	66	77	61	23
Kabui (hill tribe)	187	80	76	24	4
Tangkhul (hill tribe)	274	106	83	76	9

Tangkhul Nagas show high O followed closely by A and B groups. The comparative occurrence in respect of ABO blood groups among the groups shows that besides the Muslims who present A> O> B the same pattern of O> A> B is maintained by all others.

TABLE 5.1

ABO Gene Frequencies Among the Four Groups

Populations	*No.*	*p*	*q*	*r*
Meitei	380	0.235	0.168	0.587
Tangkhul	227	0.260	0.184	0.545
Kabui	187	0.260	0.090	0.650
Muslim	274	0.200	0.180	0.620

The frequency of the gene 'r' has been found to be high in all the four groups. The next in order of frequency is the 'p' gene while the gene for B blood group appears to be very low in all the four groups.

MN Blood Groups

A. Naga Groups from the Hills

TABLE 5.2

MN Blood Groups of Manipur Nagas

Populations	*No.*	*MM*	*MN*	*NN*	*gene frequencies* *m*	*n*
Tangkhul	132	56	48	28	0.606	0.394
Kabui	184	96	84	4	0.750	0.250

It may be seen that the frequency for the gene N is rather high in the two groups compared to other populations of Nagaland. The Nocte Nagas (Bhattacharjee, 1957) show lower N genes (0.198).

B. Communities from Valley

TABLE 5.3

MN Blood Groups Among the Valley Groups

Populations	*No.*	*MM*	*MN*	*NN*	*gene frequencies* *m*	*n*
Meitei	200	64	99	37	0.567	0.433
Muslim	200	68	105	27	0.603	0.397

The two communities from the valley have almost similar distribution of MN blood groups. The chi-square test reveals a satisfactory agreement between the observed and the expected number of phenotypes.

RH Blood Groups

TABLE 5.4

Frequency of RH Phenotypes in the Populations of Manipur

Populations	*CC Dee*	*Cc Dee*	*Cc DEe*	*cc DEe*	*cc DEE*	*cc Dee*	*Cc Dee*	*CC DEe*	*cc ddee*
Tangkhul (92)	24	32	16	2	10	2	2	2	2
Kabui (66)	34	10	16	0	4	0	2	0	0
Meitei (103)	57	20	19	4	0	0	0	2	1
Kabui (111)	61	18	23	2	1	1	1	5	0

TABLE 5.5

RH Blood Group Genes in Manipur Populations

Genes	*Tangkhul*	*Kabui*	*Meitei*	*Muslim*
Rz	2	0	2	5
R_1	56	44	76	79
R_2	16	16	19	23
r	2	0	1	0
R_0	2	0	0	1
Others (R"/E'/R_y)	14	5	5	3

'R_1' gene appears to occur in the highest frequencies in all the four studied groups. 'R_2' gene occur in second highest frequency. Among the Kabuis the gene 'Rz' is absent, while among the Tangkhuls and Meiteis this gene occurs in equal frequencies.

The Muslims have the highest frequency of this gene. The 'r' (cde) gene is absent among the Kabui and Muslims while the Tangkhul and Meitei show this with low frequencies.

When the ABO distribution of the four Manipur population groups are compared with the neighbourhood populations from Nagaland, certain important findings emerge out. It will be evident that the Meiteis and the Muslims, the two religious communities of the valley have the same ABO blood group picture (O> A> A). The Kabui follow the valley populations while the Tangkhuls present A> O> B. This pattern is however, not observed with any other Naga groups from Nagaland. All the Naga groups follow the pattern of the Meiteis. The Lushais of the Lushai hills have similarity with those of the Tangkhuls in respect of their ABO blood group distribution as will be evident from the Table 5.7.

The inter-group differences tested as a test of homogeneity by the 2 X 4 contingency table reveals the following results.

TABLE 5.6

Inter-Group Differences in Respect of ABO Blood Groups

Sl.No.	*Population groups*	*Homogeneity X 2/3 Values*	*Probability*
1.	Meitei X Tangkhul	2.986	not significant
2.	Meitei X Kabui	22.455	0.001
3.	Meitei X Muslim	14.907	0.01
4.	Tangkhul X Kabui	29.461	0.001
5.	Tangkhul X Muslim	14.096	0.01
6.	Kabui X Muslim	15.295	0.01

As will be seen that significant inter-group differences exist with a greater or lesser degree excepting the Meitei and Tangkhul which is not significant.

TABLE 5.7

Distribution of ABO Blood Groups among some of the Tribal Groups of North Eastern India

Populations	*No.*	*O*	*A*	*B*	*AB*	*gene frequencies* *p*	*q*	*r*	*Source*
Zeliang	520	271	145	91	13	0.166	0.106	0.728	Charkraborty, M 1965
Nocte	313	129	110	63	11	0.218	0.127	0.655	Bhattacharjee, 1957
Nagas	2000	673	491	651	185	0.186	0.237	0.577	Mitra, 1933
Angami	100	45	38	11	6	0.233	0.086	0.681	Bhattacharjee, 1949
Angami	165	76	64	19	6	0.241	0.079	0.680	Mitra, 1935
Nagas	140	56	47	31	6	0.213	0.143	0.644	British Assocn., 1939
Ao Naga	57	27	13	13	4	—	—	—	British Assocn., 1939
Konyak	127	58	51	13	5	0.252	0.074	0.674	-do-
Lushai	141	46	63	23	9	0.301	0.121	0.577	Mitra, 1935
Thadou Kuki	83	16	25	27	15	—	—	—	British Assocn., 1939

The Thadou Kukis have the ABO blood group distribution as B> A> O which, however, is not observed in any other population groups examined so far from this region. The r gene is as usual very high among all the Naga groups of Nagaland. The p gene varies between 0.166 (Aeliang Naga) and 0.301 (Lushai) while the q gene varies between 0.074 (Konyak) and 0.237 (Naga). It will be thus seen that excepting the Lushais as well as Thadou Kukis, who practically do not live in the Naga hills or in Nagaland, all other Naga sub-groups are characterized by high frequency of r gene. The low frequency of q gene among the Angami Nagas and the Knoyak Nagas is very interesting.

The gene B among the Nagas (British Assocn., 1939) as well as among the combined Nagas (Mitra, 1935) is quite high. This appears to be an usual serological character of this region.

TABLE 5.8

ABO Distribution among the Caste and Tribal Populations of some Plain North East Indian Origin

Populations	*No.*	*O*	*A*	*B*	*AB*	*Source*
High caste (Assam)	1691	37.31	23.59	31.30	7.68	Das *et.al.* (1979)
Low caste (Assam)	1170	38.37	25.89	28.20	7.52	-do-
Khasi (Meghalaya)	821	48.59	30.45	16.68	4.26	-do-
Rabha	600	25.50	32.50	29.83	12.17	-do-

There exists practically no difference between the high, low castes and tribals of Assam and they closely follow the patterns of Manipur population groups.

The Mongoloid element which is dominant in almost all the population groups of Assam, Nagaland and Manipur, has undergone miscegenation to a great extent. This has, however, not disturbed the ABO blood group frequency of the Muslims of Assam as well as of Manipur. It may be added that the Muslims in Assam have significant difference with those of the Hindus as well as the Khasi tribal group in respect of ABO blood group distribution. The Muslims of Manipur also behave uniformly with those of the other local population groups around Manipur and Nagaland.

Das (1974) did not observe any difference between the Assamese Muslims and the Assamese Hindus in respect of ABO blood group frequencies. The equilateral triangles have been drawn on the basis of gene frequencies (Figs. 2—5). They present the extent of homozygosity and heterozygosity of the four Manipur population groups. Among the Kabui (Fig. 2), the higher proportion of homozygous OO individuals is very

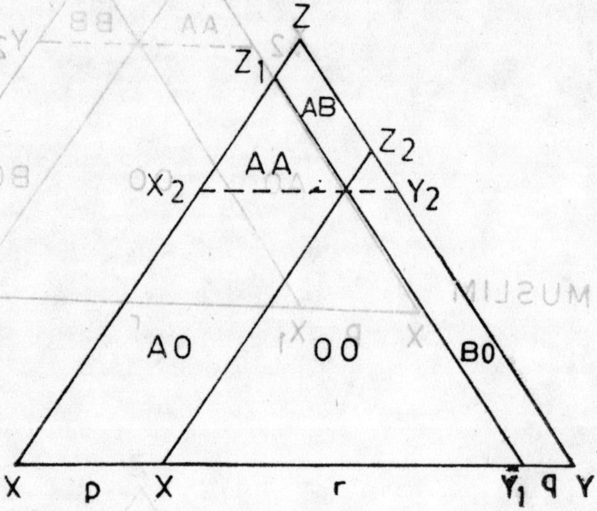

striking. The heterozygous AO has a larger proportion than the heterozygous BO. The homozygous AA is approximately equal to the combined proportion of BB and AB. The Muslims, on the other hand, have higher proportion of OO homozygous individuals, closely followed by AO and BO heterozygous individuals. The AB appears to be occupying the proportion which is equal to the combined proportion of AA and BB. (Fig. 3).

The Meitei (Fig. 4) have a larger proportion of OO than the Tangkhul Nagas (Fig. 5). There exists some difference in the proportion of AO and BO among these two groups. The AB proportion is larger in the Tangkhuls than in the Meiteis. The proportions of AA and BB are almost in equal proportion.

The close similarity in ABO blood group between the Tangkhul Nagas and the Meiteis is again confirmed indicating the ethnic homogeneity between these two groups. This was observed in other various genetic markers. The inter-group difference

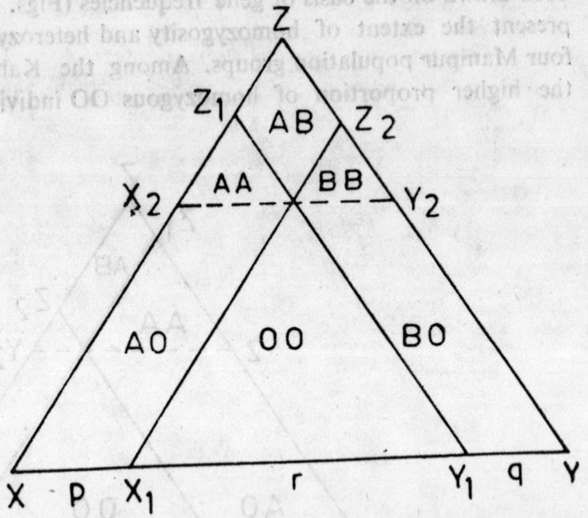

MUSLIM

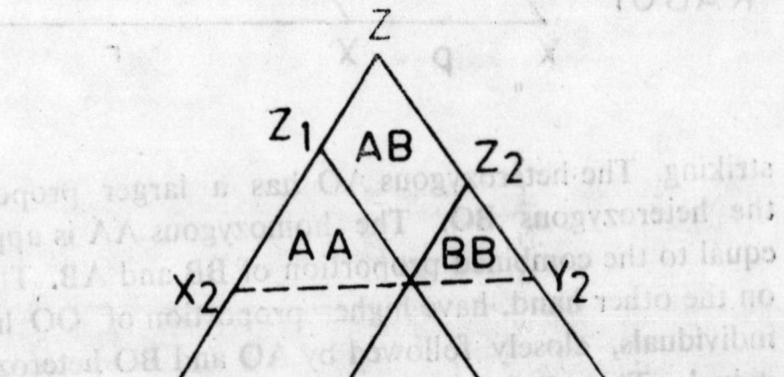

MEITEI

between the Meitei and Kabui and between the Tangkhul and Kabui are highly significant indicating their ethnic heterogeneity. Similar picture is obtained in respect of the Muslims and Kabui as well as Meitei and Muslims. The Tangkhul and the

TANGKHUL

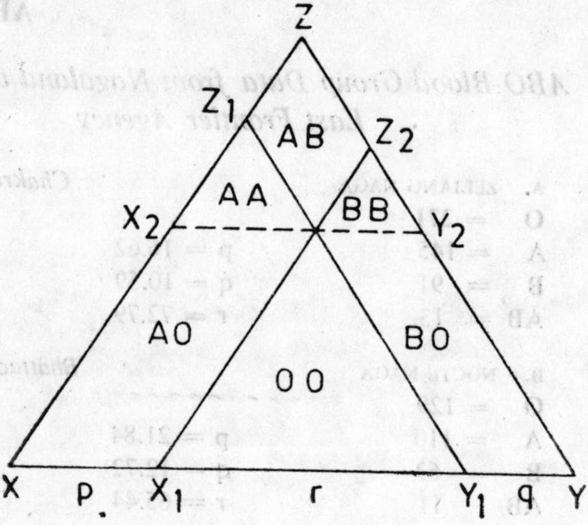

Kabui Nagas also show significant inter-group differences. Compared to the data available on ABO blood groups, we have very little informations about the MN and Rh systems of blood groups on Manipur. The two Manipuri tribal groups do not, however, show any significant variation in their phenotypic distribution.

APPENDIX

ABO Blood Group Data from Nagaland and North East Frontier Agency

A. ZELIANG NAGA — *Chakraborty : 1965*

O = 271	
A = 145	p = 16.62
B = 91	q = 10.59
AB = 13	r = 72.79

B. NOCTE NAGA — *Bhattacharjee : 1957*

O = 129	
A = 110	p = 21.84
B = 63	q = 12.72
AB = 11	r = 65.44

C. NAGAS — *Mitra : 1936*

O = 673 (33.65%)	
A = 491 (24.55%)	p = 18.61
B = 651 (32.55%)	q = 23.67
AB = 185 (9.25%)	r = 57.72

D. ANGAMI NAGA — *Bhattacharjee : 1949*

O = 45	
A = 38	p = 25.03
B = 11	q = 8.85
AB = 6	r = 66.12

E. ANGAMI NAGA — *Mitra : 1936*

O = 76 (46.06)	
A = 64 (38.79)	p = 24.13
B = 19 (11.51)	q = 7.89
AB = 6 (3.64)	r = 67.98

F. ALL NAGAS British Association of Research Committee: 1939

O = 56		
A = 47	p = 22.31	Mainly the Lhota,
B = 31	q = 14.33	Angami, Rengama
AB = 6	r = 64.36	and Sema Nagas.

G. AO NAGA *British Assocn. Res. Comm.: 1939*

O = 27
A = 13
B = 13
AB = 4

H. KONYAK NAGA *British Assocn. Res. Comm.: 1939*

O = 58	
A = 51	p = 25.21
B = 13	q = 7.35
AB = 5	r = 67.44

I. LUSHAI *Mitra: 1936*

O = 46	
A = 63	p = 30.16
B = 23	q = 12.13
AB = 9	r = 57.71

J. THADOU KUKI *British Assocn. Res. Comm.: 1939*

O = 16
A = 25
B = 27
AB = 15

References

1. Basu, R.N. (1938) : "Blood Groups among the Khasi", *Nature*, 142 : 797.
2. Bhattacharjee, P.N. (1957), "A study on ABO, MN Blood Groups and the ABH secretion in the Nocte of NEFA", *Bull. Dept.—Anth Govt. India*. VI : 1.
3. ——— (1949) : "ABO Blood Group of the Angami Naga", *Eastern Anthropologist*, 11 : 1.

4. British Research Association *(1939) : "Blood Groups among primitive peoples", *Nature*, 144 : 714.
5. Chakraborty, M. (1965) : "ABO Blood Group of the Zeliang Naga." *Man in India*, 45 : 311-15.
6. Das, B.N. (1960) : "Somatic Variations among the Hills and Plains Garos of Assam", *Man in India*, 40 : 2.
7. ——— (1974) : "Contemporary Anthropological Research in North East India", *Ed.* B.M. Das: Dibrugarh University Publications, Dibrugarh, Assam.
8. Das, F.A. (1974) : "A Comparison of the Assamese Muslims with the Assamese Hindu and the Muslims of some other states of India in respect of ABO Blood Group," in *Contemporary Anthropological Research in North-East India, Ed.* B.M. Das, Dibrugarh University Press, Dibrugarh, Assam.
9. Macfarlane, E.W.E. (1941) : "Tibetan and Bhotia Blood Group distribution", *J.A.S.B.* 7.
10. Mitra, R.N. (1936) : "Blood Groups of the Angami Naga and Lushai tribes", *Ind. Jr. Med. Res.* 63.

6

Functional Assymmetry Among Population Groups of Manipur

The best known examples of functional assymmetry studied in man are handedness, hand clasping, and folding and sitting posture. The role of heredity is most clear in the development of handedness and there are various models for the inheritance of this trait (Annett. 1964, 1967; Levy 1976). Less evident and merely partial is the genetic control of hand clasping and arm folding and this has provoked some controversy (Freire-Maia, Freire-Maia and Quelce-Salgado, 1960; Rhoades and Damon, 1973; Martin, 1973; Friere-Maia, 1976). Besides, numerous authors have noted ethnic differences in these patterns of laterality in man (Freire-Maia, Quelce-Salgado and Freire-Maia, 1958; La and Walsh, 1965; Freire-Maia and de Almeida, 1966; Hardyck *et.al.*, 1975). These reports point to the anthropological significance of the above-mentioned traits. It has been suggested that hand clasping is a more useful trait than arm folding in population comparisons (Lourie, 1972; Rhoades and Damon, 1973). Although, the frequencies of the two alternative types of both traits have been reported in number of populations throughout the world, the data from Imphal valley and the surrounding Hills of Manipur state in the extreme North East of Indian subcontinent are still scanty. Singh and Malhotra (1971) reported the distribution of hand clasping and arm folding of two Meitei Manipur groups.

The present chapter provides new data on hand clasping, arm folding, handedness and sitting posture among the four population groups of Manipur. The data was available through field investigations in 1977-78.

POPULATION GROUPS AND METHODOLOGY

The study includes 1469 apparently healthy adult individuals of both sexes aged between 18 and 30 years. The group-wise break up of the data is as follows:

1. Meitei Manipuri : M-223; F-281 Total=504
2. Tangkhul Naga ; M-159; F-110 Total=269
3. Kabui Naga : M- 80; F-256 Total=336
4. Muslims : M- 36; F-324 Total=360

Hand clasping, arm folding and sitting posture were determined by standard methods after Leguebe (1976). Handedness was determined by direct questioning of subjects, following which they were doubly checked by actual writing. The same individual was subjected to all examinations.

Results and Discussions

The percentile frequencies of hand clasping, arm folding and sitting posture of four endogamous groups appear in Table 6.0.

HAND CLASPING (TABLE 6.0)

The R trait is most frequent in both the sexes of Meitei Manipuris. The chi-square shows a homogeneous distribution (0.17). The range of variation for hand clasping is from 35.00 (Muslims) to 64.09 (Meiteis). Among the Tangkhul Nagas, the males have more R trait while the females present more L type. The two sexes present a heterogeneous distribution. The Kabui Naga males have slightly higher R trait that the L type, while the females have equal distribution of L and R traits. The two sexes present a homogeneous distribution. The combined value again follow the Meitei and Tangkhul patterns.

The Muslims present an altogether different picture. The males have equal proportion of L and R distribution while the

TABLE 6.0

Distribution of Hand Clasping in Four Manipur Populations

Populations	*Male*		*Female*		*Both*		*Chi-Square*	
	Rt.	*Lt.*	*Rt.*	*Lt.*	*Rt.*	*Lt.*	*df=1*	*P*
Meitei	145	78	178	103	323	181	0.17	ns
%	65.02	34.97	63.34	36.65	64.09	35.91		
Tangkhul	97	62	48	62	145	124	7.88	>0.01
%	61.00	38.99	43.64	56.36	53.90	46.10		
Kabui	44	36	128	128	172	164	0.60	ns
%	55.00	45.00	50.00	50.00	51.19	48.81		
Muslim	18	18	108	216	126	234	3.93	>0.05
%	50.00	50.00	33.33	66.66	35.00	65.00		

ns=not significant

females provide more L than R type. The combined M+F value also present more L trait than R trait. The two sexes present a value of 3.93 which is of low significance (P>0.05).

The results obtained by us from the Meitei Manipuris thus confirm the earlier report of Singh and Malhotra (1971).

ARM FOLDING (TABLE 6.1)

The range of variation of arm folding is from 36.14% (Kabui Nagas) to 57.35 (Muslims).

It will be evident from the combined M+F values that excepting the Muslims, all the other three groups have mostly L trait. Our result, however, does not agree with Singh and Malhotra (1971) where they had reported R trait as most frequent. While they did not observe any heterogeneity in their samples, we have observed this (X_1^2= 4.73 P>0.05). The Tangkhul Nagas, Kabui Nagas and the Muslims, however, show homogeneous distribution between the two sexes.

TABLE 6.1

Distribution of Arm Folding in Four Population Groups

Populations	Males		Females		Both		Chi-Square	
	Rt.	*Lt.*	*Rt.*	*Lt.*	*Rt.*	*Lt.*	*df=1*	*P*
Meitei	102	115	106	178	208	293	4.73	>0.05
%	47.00	53.00	37.32	62.68	41.52	59.48		
Tangkhul	78	80	47	57	125	137	0.55	ns
%	49.38	50.63	45.19	54.80	47.71	52.29		
Kabui	32	52	88	160	120	212	0.33	ns
%	38.10	61.90	35.48	64.52	36.14	63.86		
Muslim	18	12	216	162	234	174	0.12	ns
%	60.00	40.00	57.14	42.86	57.35	42.65		

ns=not significant

TABLE 6.2

Distribution of Sitting Posture in Four Population Groups

Populations	Male		Female		Both		Chi-Square	
	Rt.	*Lt.*	*Rt.*	*Lt.*	*Rt.*	*Lt.*	*df=1*	*P*
Meitei	160	56	218	53	378	109	2.71	ns
%	74.07	25.93	80.44	19.56	77.61	22.39		
Tangkhul	99	53	78	22	177	75	4.48	>0.05
%	65.13	34.87	78.00	22.00	70.23	29.77		
Kabui	52	16	192	48	244	64	0.38	ns
%	76.47	25.53	80.00	20.00	79.22	20.78		
Muslim	24	6	312	60	336	66	0.28	ns
%	80.00	20.00	83.87	16.13	83.58	16.42		

ns=not significant

Freire-Maia and Almeida (1966) reported high (60—61%) incidence of R type of hand clasping in their Japanese data, while the R type of arm folding is only about 44%. Like the Japanese, the Meitei, Tangkhul and the Kabui Nagas are also of Mongoloid origin and the values in respect of hand clasping and arm folding come very close to the Japanese data.

SITTING POSTURE (TABLE 6.2)

It will be evident from the table that excepting the Tangkhul Nagas all the other three groups show homogeneous distribution between the sexes. While in the Tangkhul Nagas, the value being 4.48 (P>0.05) the difference shows heterogeneity. The pattern R is more than the L in all the series and in both sexes.

HANDEDNESS (TABLE 6.3)

The Kabui Nagas and the Muslims do not present any left handedness. There are no sex difference between the Meiteis and the Tangkhul Nagas and they show homogeneity.

TABLE 6.3

Distribution of Handedness in Four Population Groups

	Male		*Females*		*Both*		*Chi-Square*	
Populations	*Rt.*	*Lt.*	*Rt.*	*Lt.*	*Rt.*	*Lt.*	*df=1*	*P*
Meitei	206	7	262	5	468	12	0.92	ns
%	96.71	3.29	98.12	1.87	97.50	2.50		
Tangkhul	122	7	61	1	183	8	1.47	ns
%	94.57	5.43	98.38	1.62	95.81	4.19		
Kabui	34	—	122	—	156	—	—	
%	100.00	—	100.00	—	100.00	—	—	
Muslim	10	—	124	—	124	—	—	
%	100.00	—	100.00	—	100.00	—	—	

ns=not significant

INTER GROUP DIFFERENCES (TABLE 6.4)

It will be apparent from the table, that there are no inter-group differences, so far handedness is concerned.

So far sitting posture is concerned, only weakly (P>0.05) significant values are observed between Meiteis and Tangkhul Nagas as well as between Tangkhul and Muslims.

TABLE 6.4

Inter Group Differences in Respect of Four Traits

Population Groups	*Hand Clasping*	*Arm Folding*	*Sitting Posture*	*Handedness*
Meitei & Tangkhul	7.44++	2.86	4.62+	099
Meitei & Kabui	5.96+	0.92	0.09	2.46
Meitei & Muslim	20.18+++	5.49	1.59	1.11
Tangkhul & Muslim	6.61++	1.86	4.71+	2.88
Tangkhul & Kabui	0.25	3.15	2.14	2.74
Kabui & Muslim	4.16+	6.89++	0.73	0.00

+ denotes P>0.05
++ denotes P>0.01
+++ denotes P>0.001

In the case of arm folding, low significant value is obtained between Meiteis and Muslims while the Kabui Nagas and the Muslims present significant value (P>0.01).

Hand clasping presents almost an uniform picture. Besides Tangkhul and Kabui Nagas, there are values of low to high significance between the groups as will be evident from the table.

INTER GROUP DIFFERENCES OF THREE COMBINATIONS (TABLE 6.5)

As will be seen in the table that all the values resulting out of inter group differences in respect of three combinations provide low significance to high significance.

TABLE 6.5

Inter Group Differences of Three Combinations

Population Groups	*Chi-square Values (df=7)*	*Probability*
Meitei X Tangkhul	21.22	0.01
Meitei X Kabui	27.38	0.001
Meitei X Muslim	36.44	0.001
Tangkhul X Kabui	25.97	0.001
Tangkhul X Muslim	18.71	0.01
Kabui X Muslim	15.93	0.05

FREQUENCIES OF THREE COMBINATIONS (TABLE 6.6)

Among the Meiteis, the combinations LRR and RRR have the highest frequencies. The arm folding alongwith the sitting posture behave almost uniformly in this ethnic group. The Tangkhul Nagas are characterised by RRR and LLR as the two highest frequencies closely followed by LRR combinations. The Kabui Nagas behave like that of the Tangkhul Nagas. The Muslims stand apart with RLR (28.77%) as the highest frequencies.

COMPARISON OF HAND CLASPING AND ARM FOLDING (TABLE 6.7)

Table 6.7 presents a comparison of hand clasping and arm folding between the population groups of Manipur examined so far.

The two sets of Meiteis agree very well with each other in respect of two characters, hand clasping and arm folding.

The Manipuri Brahmins who are basically Caucasoids with appreciable Mongoloid admixture, show a highly significant difference from the Meiteis. The two Naga groups also behave

TABLE 6.6

Frequencies of Different Combination of HC, AF and SP

Combinations HC	FA	SP	*Meitei*		*Tangkhul*		*Kabui*		*Muslim*	
R	R	R	(117)	23.54	(52)	20.31	(44)	13.25	(68)	23.29
R	R	L	(20)	4.02	(14)	5.47	(98)	2.41	(8)	2.74
R	L	R	(50)	10.06	(37)	14.45	(52)	15.66	(84)	28.77
R	L	L	(19)	3.82	(15)	5.86	(4)	1.20	(8)	2.74
L	R	R	(149)	29.98	(46)	17.97	(120)	36.14	(40)	13.70
L	R	L	(35)	7.04	(25)	9.76	(4)	1.20	—	—
L	L	R	(66)	13.27	(50)	19.53	(96)	28.92	(72)	24.66
L	L	L	(41)	8.24	(17)	6.64	(4)	1.20	(12)	4.10

HC=Hand Clasping
AF=Arm Folding
SP=Sitting Posture
figure in the parenthesis denote absolute numbers.

TABLE 6.7

Comparison of Hand Clasping and Arm Folding in the Two Manipur Series

Population Groups	*Hand Clasping* *Rt.*	*Lt.*	*Arm Folding* *Rt.*	*Lt.*	*Source*
Brahmin Manipur	75.22	24.78	45.62	54.38	Singh and Malhotra, 1971
Meitei Manipur	64.02	35.98	43.06	56.94	-do-
Meitei Manipur	64.09	35.91	41.52	58.48	Present study
Tangkhul Naga	53.90	46.10	47.52	58.48	-do-
Kabui Naga	51.19	48.81	36.14	63.86	-do-
Muslim	35.00	65.00	57.35	42.65	-do-

almost uniformly especially in hand clasping rather than in arm folding. The Muslims have altogether different patterns.

It, therefore, indicates that selective forces are acting differently in various populations and/or different environments. There is absolute absence of significant sex effect. Although Freira-Maia and Almeida (1966) observed that R individuals in hand clasping are the most frequent, it does not have any agreement in our data. The Muslims have more L individuals in hand clasping than R individuals while the Kabui Nagas have almost equal proportion of L and R individuals. In arm folding, L individuals are more in all the three series excepting in the Muslims where R individuals are most frequent.

When hand clasping and arm folding are considered in group differentiation, the following picture emerges:

Population Groups	*Hand Clasping*	*Arm Folding*
Brahmins	R>L	L>R
Meiteis	R>L	L>R
Tangkhul Nagas	R>L	L>R
Kabui Nagas	R=L	L>R
Muslims	L>R	R>L

References

1. Annett, M., (1964) : "A model of the inheritance of handedness and cerebral dominance", *Nature* (London) 204 : 59-60.
2. ———, (1967) : "The binomial distribution of right, mixed and left handedness", *Quarterly Jr. Exp. Psychol.* 19 : 327-333.
3. Freire-Maia, A., (1976) : "Is there any evidence for a genetic basis of hand clasping ?" *J. Hered.* 67 : 111-113.
4. Freire-Maia, A., N. Freire-Maia, and A. Quelce-Salgado, (1968) : "Genetic analysis in Russian immigrants", *Amer. J. Phy. Anth.*, 18 : 235-240.
5. Freire-Maia, N., Quelce-Salgado, A. and A. Freire-Maia, (1958) : "Hand clasping in different ethnic groups", *Hum. Biol.* 30 : 281-291.

6. Freire-Maia, A. and de Almeida, (1966) : "Hald clasping and Arm folding among African Negroes", *Hum. Bio.* 38 : 175-179.
7. Hardyck, C., R. Goldman, and L. Petrovich, (1975) : "Handedness, sex, race and age", *Hum. Biol.* 47 : 369-375.
8. Lai, L.Y.C. and Walsh, (1965) : "The patterns of hand clasping in different ethnic groups", *Hum. Biol.* 37 : 312-319.
9. Leguebe, A., (1967) : "Hand clasping : etude anthropologique et genetique", *Bulletin de la Societe Royale Belge d' Anthropologie Prehistorique* 78 : 81-107.
10. Lourie, J.A. (1972) : "Hand clasping and Arm folding among Middle Eastern Jews in Isreal", *Hum. Biol.* 44 : 3.
11. Levy, J., (1976) : "A review of evidence for a genetic component in the determination of handedness," *Behavioural Genetics*, 6 : 429-453.
12. Martin, N.G. (1975) : "No evidence for a genetic basis of tongue-rolling and hand clasping," *J. Hered.* 66 : 179-180.
13. Rhoads, J.G. and A. Damon, (1973): ' Some genetic traits in Solomon Island populations", *Amer. J. Phys. Anth.* 39 : 179-183.
14. Singh, N.R. and K.C. Malhotra, (1971) : "Hand clasping and Arm folding among the Manipuris (India)", *Hum. Hered.* 21 : 203-207.

7

Ear Lobe and Tongue Rolling Among the Manipur Peoples

The type of attachment of ear lobe in man is an easily observable genetic trait, presumably having no adaptive value. The genetics of this trait is yet to be fully understood. This trait, therefore, cannot at present be satisfactorily utilized in exact linkage studies in man. However, the ear lobe attachment has been profitably used in population studies as a taxonomic character.

Data on this trait is now available for a large number of Indian population. Singh and Malhotra (1970) have reported the ear lobe frequencies among the Brahmins and the Meiteis of Manipur. The present chapter deals with the new data on ear lobe and tongue rolling from Manipur population groups. Observations on ear lobe types were made on 1326 adult individuals from both sexes of four endogamous population groups from Manipur. The types of ear lobe attachment have been divided into three well defined categories—free, intermediate and attached. While examining the individuals, the ear lobes were slightly drawn apart to see the extent of attachment. This was done in order to avoid any observational error, particularly in the case of small ear lobes.

Results and Discussion

Table 7.0 describes the frequencies of the three types of ear lobes in four population groups of Manipur.

TABLE 7.0

Frequencies of the types of ear lobe attachment in four population groups of Manipur

Population	No.	*Ear Lobe/Attachment*					
		Free	%	*Inter*	%	*Attached*	%
Males							
1. Meitei	222	40	18.01	66	29.73	116	52.26
2. Tangkhul	160	19	11.88	34	21.24	107	66.88
3. Kabui	104	26	25.00	30	28.85	48	46.15
4. Muslim	36	10	27.78	10	27.78	16	44.44
Heterogeneity X^2_6 =2.97 not significant							
Females							
1. Meitei	277	42	15.16	100	36.10	135	48.74
2. Tangkhul	103	15	14.56	35	33.98	53	51.46
3. Kabui	228	66	28.95	50	21.92	112	49.12
4. Muslim	196	60	30.61	48	24.49	88	44.90

Heterogeneity X^2_6 =2.23 not significant

It appears from the above table that there is no sex-difference in this trait among the four population groups. The combined values of all males vs. all females, however, do not show heterogeneity (X^2_2=4.6 P less than 0.05).

From the results above, it appears that this trait occurs as a polygenic variable in man. This may, therefore, be profitably used as a taxonomic character.

Table 7.1 explains the inter-group differences, on the basis of two types only, attached and free.

TABLE 7.1

Inter-group differences in respect of ear lobe types

Population Groups			*Chi-square Value (df=1)*	*Probability (P)*
1. Meitei	X	Tangkhul	7.56	0.01
2. Meitei	X	Kubui	0.04	ns
3. Meitei	X	Muslim	23.69	0.001
4. Tangkhul	X	Kabui	1.04	—
5. Tangkhul	X	Muslim	38.31	0.001
6. Kabui	X	Muslim	14.08	0.001

There exist significant inter-group differences, as is evident from Table 7.1 above.

As expected, the Muslims, the Kabui and the Tangkhuls differ from each other with regards to the ear lobe types.

Table 7.2 presents a comparison of two series of data available so far from Manipur (Singh and Malhotra, 1970).

It may be stated that both sexes of the four groups show high occurrence of attached ear lobes. The Brahmins of Manipur are basically caucasoid though with some Mongoloid admixture, while the Meiteis are fully Mongoloid. It will be evident from the table 7.2 that both Meitei and Tangkhul present attached > inter > free; while the Kabui and the Muslim present attached > free > intermediate type of ear lobe. The inter-group differences, have, therefore, highly significant values when Muslims were examined against the three groups. The Meitei X Tangkhul show significant result, because of the variation in the distribution of the three types of ear lobes.

Mahapatra and Das (1973) have studied the ear lobe types

of some Assam population groups. The tribal groups of Assam namely, the Lalung, Miri, Garo and the Kacharis show high frequency of attached ear lobe compared to the Muslims (18.5%) and other Hindu castes (average 29.6%).

TABLE 7.2

Comparative occurrence of ear lobe types among population groups of Manipur

Population groups	*No.*	*Attached*	*%*	*Free*	*%*	*Inter*	*%*	
1. Brahmin	116	26	22.35	45	38.81	45	38.81	*
2. Meitei	217	120	55.50	34	15.58	63	28.90	*
3. Meitei	499	251	50.30	82	16.43	166	33.27	+
4. Tangkhul	263	160	60.83	34	12.93	69	26.24	+
5. Kabui	332	160	48.19	92	27.71	80	24.10	+
6. Muslim	232	104	44.83	70	30.17	58	25.00	+

* Singh and Malhotra, 1970
+present study

When the attached and free ear lobe types of the two sexes of the Muslims are examined there, exists a weak X^2 value (4.44 $P>0.05$).

It may, therefore, be stated that the attached type of ear lobe is the characteristic of the Manipur population groups. This type is also noticed in the tribal groups in a higher frequency. The minor differences as well as variation between the groups in terms of the distribution of the three ear lobe types may be ascribed to the degree of admixture between them.

Tongue Rolling

About 1306 individuals belonging to four ethnic groups were examined in terms of tongue rolling—positive or negative.

It appears from Table 7.3 that there exists no striking sex difference among the population groups examined. The Kabui and the Muslims show X^2 values which are less than 0.01.

TABLE 7.3

Distribution of Tongue rolling in four groups

Population	*Sex*	*+ive*	*—ve*	X^2_1
1. Meitei	Male	122	90	0 97
	Female	147	129	
	Combined	269	219	
2. Tangkhul	Male	82	70	0.26
	Female	54	52	
	Combined	136	132	
3. Kabui	Male	48	24	4.74
	Female	120	108	
	Combined	168	132	
4. Muslim	Male	16	4	5.48
	Female	132	108	
	Combined	148	112	

It is difficult to interpret the distribution pattern of the tongue rolling since there exists no difference between the occurrence of presence and absence of this behaviour.

There exists no inter-group differences as will be evident from Table 7.4.

The positive or presence of tongue rolling behaviour varies between 58% and 54% while the absence type varies from 42% to 46%. Apparently, it appears that the presence type of behaviour is more although statistically this is not significant.

TABLE 7.4

Inter-group difference in the four groups

Population Groups	X^2_1 Value	Probability
1. Meitei X Tangkhul	0.37	ns
2. Meitei X Kabui	0.[illegible]5	ns
3. Meitei X Muslim	0.07	ns
4. Tangkhul X Kabui	0.27	ns
5. Tangkhul X Muslim	0.29	ns
6. Kabui X Muslim	0.00	ns

References

1. Singh, N.R. and K.C. Malhotra (1970) : "Ear Lobe among the Brahmins and Meiteis of Manipur", *Bull. Deccan College Res. Inst.* XXX: 165-168.
2. Deka Mahapatra, U. and P.B. Das (1973) : "Ear Lobe among the Assam populations", *Man in India,* 53 : 3 : 256-261.

8

Colour Blindness and Taste Blindness

An attempt has been made in this chapter to study the two behavioural genetical traits among the four population groups of Manipur—three from the valley and one from the hills.

COLOUR BLINDNESS

The following population groups alongwith their respective sample size were examined:

1. Meitei (200)
2. Tangkhul Naga (300)
3. Kabui Naga (242)
4. Muslim (200)

The examination was conducted in the field in various schools and the range of age was from 15 to 20 years. The trait was examined with the help of the Ishihara plates (1962) in a room with sufficient daylight. The plate No. 1 was shown first to all the subjects and the order of the plates was changed now and then while testing. The answers were verified several times more than one investigation and finally these were recorded on the proforma.

Results

The incidence of colour blindness among the four ethnic groups is summarized in Table 8.0.

TABLE 8.0

Incidence of colour blindness among the four groups

Population Group	*No.*	*Colour Blind*	*% of Colour Blind*
1. Meitei	200	9	4.50
2. Tangkhul	300	8	2.50
3. Kabui	242	12	4.96
4. Muslim	200	10	5.00

Out of the total of 942 persons tested, 39 were found to have defective colour vision. The average percentage is 4.13. The Tangkhul Nagas show the lowest frequency of colour blindness while the Muslims have the lowest value. The values were treated statistically to see the inter-group differences. The total individuals who are 39 in number, are distributed into Protan and Deutan with strong and mild degrees in the following way :

Population	*Protan*		*Deutan*		*Total*
	strong	mild	strong	mild	
Meitei	3	2	2	2	9
Tangkhul	2	2	2	2	8
Kabui	3	4	3	2	12
Muslim	3	2	3	2	10

Table 8.1 explains the inter-group differences.

The chi-square values reveal no significant differences. The frequency distribution of the subtypes of red-green colour blindness does not show any difference.

TABLE 8.1

Values of chi-square test for inter-group differences

Population Groups	X^2_1 *Value*	*Probability*
Meitei X Muslim	0.22	ns
Meitei X Kabui	0.204	ns
Meitei X Tangkhul	1.006	ns
Kabui X Tangkhul	1.888	ns
Muslim X Kabui	0	
Muslim X Tangkhul	2.182	ns

TABLE 8.2

Colour blindness in some North East Indian Population Groups

Population	*% of Occurence in males*	*Source*
1. Riang	1.54	Kumar *et. al.*, 1961
2. Khasi	3.87	Mukherjee, 1963
3. Lushai	1.78	Mukherjee, 1963
4. Upper Assam Hindus	3.61	Shrivastava, (unpub.)
5. Upper Assam Muslims	1.41	Shrivastava, (unpub.)
6. Lower Assam Hindus	3.71	Das and Das (1973)
7. Lower Assam Muslims	2.56	Das and Das (1973)

In the absence of any data on Manipur population groups, we are unable to compare our data with those of other Manipur population groups. Das and Das (1973) have examined the

Assamese Hindus and Muslims in respect of colour blindness. The $X^2{}_1$ value is, however, not significant (0.393) P—0.70. The Assamese Muslims have a frequency of 2.56% of colour defect. Deutans (green blind) is higher than that of the protans (red blind) in both the groups. In our sample, we have found 21 Protans and 18 Deutans.

The colour vision defect in some North East Indian population groups are mentioned in Table 8.2.

It may be seen from the table that the Assam Hindus have always high frequency of colour blindness compared to the Muslims. Similarly among the tribes, the Khasis inhabiting Meghalaya presents 3.87 per cent belonging to colour blindness group.

The Riangs, the Khasis and the Lushais are Mongoloid tribes while the Assamese Hindus and Muslims are caucasoid groups. The Riang and the Lushais show low frequencies of colour blindness in comparison to that of the Assamese Hindus.

Mukherjee (1963) stated "the Riang have ethnic and geographical propinquity with the Lushais."

Compared to other North East Indian population groups, examined so far, the population groups belonging to Manipur show high colour blindness value.

TASTE BLINDNESS

Two different types of investigations related to taste blindness were undertaken. The first type of study was the determination

TABLE 8.3

Distribution of PTC threshold among the Muslims of Imphal

Threshold No.	*0*	*1*	*2*	*3*	*4*	*5*	*6*	*7*	*8*	*9*	*10*	*11*	*12*	*13*	*14*
Males	53	3	9	9	11	17	28	18	5	4	1	—	—	—	—
Females	31	2	4	5	13	21	13	20	4	3	—	—	—	—	—
M+F	84	5	13	14	24	38	41	38	9	7	1	—	—	—	—

Chi-square between the two sexes=0.144

TABLE 8.4

Frequency of taster and non-taster among the Muslims

Population	*No.*	*Taster*	%	*Non-taster*	%	*T*	*t*
Muslims	274	190	69.34	84	30.66	0.446	0.554

of PTC taste threshold after the method of Harris and Kalmus (1949). From Imphal, 274 Muslims were subjected to threshold determination.

The percentage frequency and the allele frequencies of tasters and non-tasters have been calculated after noting that the anti-mode is at solution number 1 in both the sexes. It will be seen from Table 8.3 that the males have threshold upto 10 while the females have upto 9. There appears no statistical difference between the two sexes. The Muslims are 69.34 tasters.

The other type of test was conducted without threshold, where the subjects were classified into taster and non-taster only with the help of the stock solution.

TABLE 8.5

The taster and non-taster distribution in Manipur populations

Population	*No.*	*Phenotype Distribution*		*Gene Frequencies*	
		taster	*non-taster*	*T*	*t*
1. Meitei	271	234	37	63.05	36.95
2. Tangkhul	271	230	41	60.80	39.20
3. Kabui	216	178	38	57.80	42.20
4. Muslim	200	168	32	39.54	60.46

Since the distribution of the taster and the non-taster phenotypes in the two sexes are found to be homogeneous as indicated by the chi-square test, we have calculated the gene frequencies of the pooled sample.

Excepting the Muslims who present 0.605 (60.46%) of non-taster gene, the other three population groups show almost uniform pattern. The t gene varies in these three samples from 36.95 to 42.20 per cent.

The inter-group differences are examined in the table below.

TABLE 8.6

Inter-group differences in respect of taste sensitivity

Population	X^2_1 value	Probability
1. Meitei X Muslim	0.627	ns
2. Meitei X Kabui	1.599	ns
3. Meitei X Tangkhul	0.240	ns
4. Muslim X Kabui	0.275	ns
5. Muslim X Tangkhul	0.066	ns
6. Kabui X Tangkhul	0.552	ns

The inter-group differences do not exhibit any statistically significant result.

In the absence of other Manipur data, an attempt has been made to compare the present data with those of other North East Indian data (Table 8.7).

It will be apparent from Table 8.7 that the frequency of 't' gene in Manipur presents an uniform picture. The Riangs and the Nokte Nagas are in agreement with the Manipur data indicating that the Mongoloids of North East India have the 't' frequency varying from 13.66 to 17.58 per cent. Similarly the caucasoids of Assam, the Brahmins, Kalita and the Muslims present an uniform picture and the values are higher than those of the Manipur populations. The Khasis, a Mongoloid group of Meghalaya, however, present a very low 't' gene. The high

TABLE 8.7

"t" gene in North East India

	Population	*No.*	*"t" gene %*	*Source*
1.	Khasi	317	2.18	Miki *et. al.* (1960)
2.	Nokte Nagas	271	13.66	Kumar (1955)
3.	Riang	401	16.21	Kumar & Sastry, 1961
4.	Brahmins (Assam)	189	39.1	Das, 1971
5.	Kalita	310	35.8	Das, 1971
6.	Muslim (Assam)	275	42.6	Das, 1971
7.	Meitei	271	13.66	Present study
8.	Muslim	200	16.00	-do-
9.	Tangkhul	271	15.13	-do-
10.	Kabui	216	17.58	-do-

castes of Assam have an average of 25.08% of non-taster gene compared to the low castes of Assam (18.83%). The X^2_1 value is, however, not significant.

Reference

1. Das, B.M. (1971) : "Taste sensitivity to phenylthiourea among the Khasi of North-East India", *Proc. International Symposium on Human Genetics*, Andhra University, Waltair.
2. ———, and P.B. Das (1973) : "Frequency of colour blindness in five groups of Assam", *Men in India*, 53 : 4 : 363-367.
3. Harris, H & H. Kalmus (1949) : "The measurement on taste sensitivity of phenylthiourea", *Annals of Eugenics*, 15 : 24-31.
4. Ishihara, S. (1962) : *Tests for Colour Blindness*, Tokyo, Japan.

5. Kumar, N & D.B. Sastry (1961) : "A genetic survey among the Riang—a mongoloid tribe of Tripura", *Z. Morph. Anth.* 51 : 3.
6. Kumar, N. (1955) : "Taste, Middle phalangeal hair and occipital hair shorls among Nocte Naga", *Bull. Dept. Anthropology.* Govt. Ind. IV : 2 : 61-67.
7. Miki, T. *et. al.* (1960) : "On the distribution of secretor character of the Lepchas and the Khasis", *Proc. Jap. Acad.* 36 : 4.
8. Mukherjee, D.P. (1963) : "Variation in defective colour vision among some tribes of Assam", *Bull. Anth. Survey Ind.* XII : 1-2 : 39-42.
.Shrivastava, R.P. : "Frequency of Colour Blindness in Assam" (unpublished).

9

Genetic Cline in Human Cerumen Types in North East India

Introduction

Adachi (1937) was the first to point out the ethnic differences in connection with the human cerumen types. His data on the incidence of cerumen type as well as axillary odour from various parts of Japan and neighbouring populations were of great ethnic value. Since then, several papers have appeared on this trait. Matsunaga (1962) and Omoto (1975) compiled a large reference on Mongoloid population groups of South East Asia and Pacific population, while Kalmus et. al. (1964) and Petrakis et. al. (1967) reported the data on cerumen types on the American Indians. From these data it is, however, seen that the frequencies ranged from a very high frequency of the allele for dry cerumen in northern Chinese, Koreans, Mongols and Japanese through intermediate values of South East Asia and Micronesia to a low frequency in Caucasians and Negroes. The information available from Indian on human cerumen types is still very limited.

Das (1977) studied the cerumen types in Assam while Chakravartti and Chakravartti (1978) proposed a three-tier ratio with regard to the gene frequencies of cerumen types among the Indian population groups. Their investigation was limited to the five North Indian population groups as well as five South Indian population groups. It was found that the

Muslims of both, North and South India, have 30:70 (wet: dry) ratio; the Christians and the Hindus of South India present a 40:60 (wet: dry) ratio while the Sikhs and the Hindus of North India have 25:75 (wet: dry) ratio. The study of Das (1977) reveals, that at one extreme the Mongoloid Kuki-Hmar tribes showing high preponderance of the dry type and at the other end the caucasoid high caste population of Assam having the wet type in high percentage.

The genetic dimorphism and distribution of human cerumen types among the various Mongoloid population groups of Assam, Manipur and Nagaland in the North-Eastern region of the sub-continent suggest that dry allele is significantly predominant. The present communication attempts to find out whether there is an evidence for a genetic cline in respect of ear wax type in this part of the country especially where the dry allele is singificantly present than that of the wet.

MATERIALS AND METHOD

The present search for a genetic cline is based on a material of 4290 individuals (both sexes) from Manipur, Nagaland and border areas of Mizoram. For comparison, the data of Das (1977) from Assam has been accounted. The investigation was carried out between January and November, 1978. The type of cerumen was determined by simple otoscopic examination of both the ear canals with the help of a small metal stick.

Results and Discussion

The population-wise break up of the data from the North-Eastern India is presented in Table 1.

It will be evident from Table 9.1 that the two Manipuri population groups from Imphal valley have very high incidence of dry allele. These two Hindu caste groups are, according to Waddel (1901) closely linked to the Indo-Chinese ethnic affiliation. The striking similarity of the high frequency of dry cerumen among the two groups further suggest their ethnic homogeniety. They come closer to the Japanese and Mongols of Matsunaga (1962).

TABLE 9.1

Human Cerumen Types in North-Eastern India

Sl. No.	*Population Groups*	*n*	*Dry Allele*	*Source*
	Manipur			
1.	Meitei Manipur (Caste)	642	0.952	present study
2.	Vishnupriya (Caste)	432	0.979	present study
3.	Tangkhul Nagas (Tribe)	492	0.972	present study
4.	Kabui Nagas (Tribe)	490	0.906	present study
5.	Chothe Kuki (Tribe)	302	0.888	present study
6.	Thadou Kuki (Tribe)	314	0.806	present study
7.	Hmar Kuki (Tribe)	342	0.896	present study
8.	Veipei Kuki (Tribe)	276	0.890	present study
9.	Mao Nagas (Tribe)	376	0.914	present study
	Nagaland			
10.	Ao Nagas (Tribe)	282	0.926	present study
11.	Angami Nagas (Tribe)	342	0.937	present study
	Assam			
12.	Brahmin (Caste)	119	0.542	Das (1977)
13.	Kalita (Caste)	224	0.526	Das (1977)
14.	Kayastha (Caste)	47	0.526	Das (1977)
15.	Kaibarta (Caste)	69	0.648	Das (1977)
16.	Kock (Tribe)	54	0.720	Das (1977)
17.	Muslim (Caste)	68	0.594	Das (1977)
18.	Bodo (Tribe)	105	0.780	Das (1977)
19.	Kuki (Tribe)	48	0.902	Das (1977)
20.	Hmar (Tribe)	67	0.906	Das (1977)
21.	Hmar & Kuki (Tribe)	110	0.904	Das (1977)

Ethnically, the tribes of Manipur may be broadly classified into two ethnic groups: (a) the Nagas and (b) the Kukis. The Naga tribes of Manipur have ethnic affinities with the people of Nagaland while the non-Naga tribes belonging to the Kuki-Chin family, have affinity with the Chins of Burma and the Mizos of Mizoram. It will be apperent from Table 9.1 that the five Naga groups—three from Manipur and two from Nagaland—have very high frequency of dry cerumen. Strikingly, the two Naga groups have almost similar gene frequency while the three Manipur Nagas, the Tangkhul Naga group living in the eastern part of Manipur at an altitude of 6500 ft. have the highest gene frequency of the dry allele; the Mao and the Kabui Naga groups, settled in the valley, have comparatively less value than that of the Tangkhul Nagas. The test of singificance between the groups shows the following results:

(a) between the two Manipuri groups = 2.55 (P less than 0.05)
(b) between the three Naga groups = 9.31 (P more than 0.01)
(c) between the Manipur & the Nagas = 4.63 (P more than 0.05)

The presence of fairly reasonable percentage of wet cerumen among the Kabui Nagas as well as the Mao Nagas living in the valley of Imphal has contributed a 0.01 per cent significance between the three groups of Manipur. The distribution of almost equal frequencies of dry and wet cerumen types has provided an ethnic homogeniety between the two Manipuri groups.

The Kuki-Chin or non-Naga tribes of Manipur are more than a dozen in number who have essentially dry allele and the gene frequencies are almost uniform. The homogeniety chi-square presents a value of 8.26. A comparison between the Naga groups and the non-Naga groups is presented below:

(a) All Kukis = 8.26 not significant
(b) All Kukis vs. all Nagas =12.11 P>0.001
(c) All Nagas vs. all Manipuris = 4.63 P | 0.05
(d) All Kukis vs. all Manipuris =28.64 P | 0.001
(e) Manipuri Nagas vs. Nagaland Nagas = 0.12 not significant

The Assam population groups present a picture with two

extremes. The high caste groups of Assam have 50 : 50 (wet: dry) ratio while the Kuki, Hmar as well as the Kuki plus Hmar combined group, on the other hand, show high preponderance of the dry type. They even have more values than those of the Manipuri Kukis. It, therefore, appears that the high preponderance of the dry type is the characteristic of all the Kuki-Chin groups of the North-Eastern India. Although, the Bodos are of Mongoloid origin yet they have a very low value—probably due to their mixture with the non-Mongoloid population groups of the plains. The dry allele is found significantly in the Middle East and in South East Asia than in American Whites and Negro population groups (Table 9.2).

The dry allele increases from the South (Kukis) and the highest value is found among the Kabui Nagas in the North-West. Similarly, the value goes on increasing from the North (Nagaland) to the Central valley of Imphal with high frequencies among the Manipuris. The evidence for a genetic cline in ear wax types in North-Eastern India especially with regards to the presence of dry allele in a high frequency than that of the type, therefore, seems to be appropriate (Fig. 1).

Summary

The genetic dimorphism and distribution of human cerumen types among the various Mongoloid groups of Assam, Manipur and Nagaland in the North-Eastern region of India suggest that dry allele is significantly prodominant than that of the wet allele. A search for a genetic cline in respect of cerumen types has been made on a data of 4290 individuals of both sexes from Manipur and Nagaland. It appears that the high incidence of dry allele is the characteristic of all the population groups of this region studied so far. The Kukis occupying the South-West of Manipur have a comparatively low value of dry allele than those of the Nagas as well as the Manipuris, but presenting a genetic homogeniety among themselves. The two Manipuris from the central valley have again very close similarity not only between them but also with those of the Manipuri and Nagaland Nagas. In Assam the Caucasoid castes and the Mongoloid tribes lived side by side in a course of time; exchange of

TABLE 9.2

Dry Gene Frequencies of Cerumen Types in Some Indian Ethnic Groups

Sl. No.	*Ethnic Groups*	*Dry Allele (q)*	*Source*
1.	Northern Hindus (high castes)	0.7600	Chakravartti & Chakravartti, 1978
2.	Sikhs	0.7700	Chakravartti & Chakravartti, 1978
3.	Northern Muslims	0.7050	Chakravartti & Chakravartti, 1978
4.	Southern Hindu (high castes)	0.6040	Chakravartti & Chakravartti, 1978
5.	Southern Muslims	0.6940	Chakravartti & Chakravartti, 1978
6.	Southern Christians	0.5920	Chakravartti & Chakravartti, 1978
7.	North-Eastern Hindus (high castes)	0.5314	Das, 1977
8.	North-Eastern Nagas (tribes)	0.9310	Present study
9.	Kukis (tribes; Assam)	0.9042	Das, 1977
10.	Muslims (Assam)	0.5941	Das, 1977
11.	Manipuris (castes; Manipur)	0.9655	Present study
12.	Kukis (tribes; Manipur)	0.8700	Present study

gene took place between the tribals and the castes. Even then, a gradient could be drawn in respect of their dry genes. The variation in the frequencies of this gene in Assam may be attributed to the admixture. Among the Mongoloids of North-Eastern India, there exists regional geographical variation. In Manipur, on the other hand, the admixture was more with the local neighbourhood Nagas and that is why probably the genetic cline could be evident. Further search for this gene among the Mizos and the Burmeese may perhaps help us more in testifying the hypothesis.

References

1. Adachi, B. (1937) : "Das Ohrenschmalz alse Rassenmerkmal und der Rassengeruch (Achsel-geruch) nebst dem Rassenunterschied der Schwei Bdrtisen",—*Z. Rassenkde.*, 6 : 237.
2. Chakravartti, M.R. & Rama Chakravartti, (1978) : Cerumen types in some Indian population groups",—*Z. Morph. Anthrop.* 69 : 79-85.
3. Das, B.M. (1977) : "Inheritance of cerumen type and its importance", —*Man in India* 57 : 4.
4. Kalmus, H., A.L., De Garrary, W. Rodarte & L. Corbo (1964) : cited from *Petrakis* et. al. (1967).
5. Matsunaga, E. (1962) : The dimorphism in human normal cerumen. —*Ann. Hum. Genet.* 25 : 273.
6. Omoto, K. (1975) : "Anthropological and Genetic Studies on the Japanese"—*JIBP Synthesis,* Vol. II. Univ. of Tokyo Press, Tokyo, Japan.
7. Petrakis, N.L., K. Molohon, & D. J. Tepeer, (1967) : "Cerumen in American Indians; Genetic implications of sticky and dry types", —*Science* 158 : 1192.
8. Waddel, L.A. (1901) : *The tribes of Brahmaputra valley*, Reprinted in 1975 by Sanskaran Prakashak, Delhi.

10

Ethnic Bearing on the Population Groups of Manipur with Reference to North-East India

Research in physical anthropology in North-East India was aptly initiated by Waddle (1900) when he measured about 564 individuals belonging to 33 different population groups of the Brahmaputra valley. His monumental research presenting ethnological notes to each population group and accompanied by their respective morphometric characters is still considered as one of the excellent monographs in this area of research.

After a lapse of about a decade, Brown and Kemp (1915) measured 84 males and 10 females, belonging to the Abor tribal population group. Hutton (1921) collected anthropometric materials from 234 individuals belonging to seven Naga tribal groups. Dixon (1922) measured 25 Khasi, 8 Sema Naga and 10 Angami Naga. He made an attempt to throw some light on the racial history of Assam. Haddon (1924) collected the four series of data on the basis of which he was able to identify seven ethnic strains who have entered into the peoples of Assam. Haddon also dealt separately in respect of Assam. Buxton (1925) almost followed Haddon in this respect, in describing the peoples of Assam and North-East India.

The morphometric characters of the Kukis were studied by Shaw (1929) and Das (1945). Each one of them measured 60 male individuals from these two tribes. The Khasi tribe was investigated by Guha (1921), Ray Chaudhuri (1935) and Bowels (1943). Singh (1935) added a new series of anthropometric data on the Sema Naga and Angami Naga. Dutt (1938) undertook a comparative anthropometric study of some of the caste groups of Assam.

Guha and Basu (1931) published a report on the human relics recovered by the Naga hills expedition for the abolition of human sacrifice during 1926-27. They found two racial types among the Naga crania, which were termed as Group I and Group II. The latter was identified to be Australoid and the former as Mongoloid.

These were the anthropometric studies carried out till the independence period. Now, while taking a stock-taking of the blood group studies during this period in this region, we may add three series of ABO blood group data collected by Mitra (1936). He examined the Assamese of the plains, the Angami Nagas and the Lushai tribals from the hills.

Basu (1938) examined the two series of Khasis while Macfarlane (1941) examined the Tibetan and Bhotia blood samples alongwith one set of the Khasis. Tne British Association of Research (1939) reported on the two groups of the Nagas from ABO blood groups point of view.

The post-independent research in physical anthropology was mainly initiated by Das and his colleagues (1960).

In 1953, Pakrasi described the physical characters of the Mikirs of Assam. Das (1960) studied the Garos. Phookan (1961) described the anthropometric characters of the Jharua Kachari and Sonowal Kachari (1962). Das (1962) examined the Bodo Kachari and the Rajbansi. Olivier (1958) dealt with the physical anthropology of the Nagas. Sharma (1961) measuren the Miris and Deuris (1962) while Das and Bhagabati (1963) dealt with the physical features of the Khamyang. Hazarika (1973) measured the Ahoms.

Das (1960) published his book "Ethnic affinities of the Rabhas" and concluded that in many features the Rabhas resemble the linguistically related Garo. In his "Anthropometric of the tribal groups of Assam, India" Das (1970) grouped the mongoloid tribes under the Khasi, the Bodo, Lushai-Kuki, the Naga and the other and showed their ethnic relation in respect of anthropometry. These mongoloid tribes were compared with other Asiatic mongoloids. Finally, an attempt was made to find out the ethnic elements which have entered into the tribal population of Assam.

A deviation was, however, observed when the problem of micro-evolution was dealt with by Das (1964) on his study on the Hill Garo and Plain Garo. As in the case of anthropometry, dermatoglyphics of the various populations have also been reported by different workers from time to time.

Bhattacharjee (1955) studied the Abors; Kumar (1954) studied the Galong of NEFA; Das and Bhagabati (1962:1963) studied the Khamyang; Shrivastava studied the Miris (1969); Das and Deka (1958:1960) studied the Suri caste; Das and Das (1965) examined the Kalita; Dutt (1973) studied the Singpho; Das and Banerjee (1972:1973) studies the Ahoms "Deka" and Bora (1973) studied the Miris while Sharma (1962) studied the Miris. A team from Japan under the supervision of Miris (1960) worked on the Khasis. Das (1974) outlined his studied on the finger and palm prints of many of the Bodo populations like Garo, Rabha, Kachari, Hajing and Rajbanshi besides undertaking a detailed study of the Khasis.

Chakravartti (1963) collected a large data on dermatoglyphics from Assam, Nagaland and Manipur. In 1961, Chakravartti and Mukherjee made an attempt to study populations of Assam in respect of dermatoglyphics. Very little research on demography and fertility has so far been done in this part of the country. Mention may be made in this connection of Deka (1970) who worked on the Khasis; Das (1978) on the Rabhas; Das and Das (1967) and Rakshit (1960) and Gogoi (1972) reported about the menarche and menopause among the Assamese girls and the Ahom women.

As stated earlier that data on ABO blood groups have mostly come from Khasi (Basu, 1938); Macfarlane (1941);Wiki et.al. (1960); Flatz et.al. (1972) and Das (1968). The Naga samples were examined by Mitra (1936); Bhattacharjee (1957) and British Association (1939) while the Lushais were examined by Mitra (1936) and Thado Kukis by the British Association (1936). The Rabhas were examined by Das (1960) and the Garos by Mazumdar (1950). Gupta (1958), Kumar (1958) and

Kumar and Sastry (1961) examined the Tripura tribal groups. Of the Arunachal tribes, the Galong were studied by Kumar (1954), Bhattacharjee (1954:1957). Serological study in North-East India was mainly confined to the determination of the percentage occurrence of the ABO blood gene and the comparisons of the populations in order to assess the ethnic affiliation. Bhattacharjee (1957) studied the MN blood type and the ABH secretion of the Nocte of Arunachal Pradesh. Kumar and Sastry (1961) made a genetic survey of the Riangs of Tripura.

Similarly, the Khasis were tested for the PTC by Das (1971). It was followed by Miki et. al. (1960), Phookan (1974) and others. In 1971, a genetic survey in three populations of Assam was undertaken by Flatz et. al. (1972) and Goedde et. al. (1972). This study indicated the importance of genetic studies of the peoples of this region with the help of various genetic markers.

Compared to Assam and its neighbourhood areas, very little research has been attempted in the state of Manipur either on the valley people or on the hill populations. It was only in 1960-61, M.R. Chakravartti and Rama Chakravartti undertook an extensive dermatoglyphic survey of the tribes and castes of Manipur. Singh and Malhotra (1970) examined two genetic traits on the two sections of the Manipuris. In 1977, therefore, a detailed research plan was maue to work on the four population groups of Manipur. These four groups were chosen, because they represent two religious communities of the valley and two dominant Naga populations from the hills. Unfortunately, due to various difficulties, data on the Kukis on a number of genetic markers could not be incorporated. It is expected that in near future we will add their results. We propose to offer a summary of the investigation undertaken on the four population groups of Manipur.

1. REPRODUCTIVE PERFORMANCE

Tables 2.0—2.3 describes the age-wise distribution of ever-married women, mean number of living and dead children and the respective proportion of miscarriages and abortion in the four groups. There is a sudden increase in the fertility rate in the middle age group (30-44 years) among the Meitei Manipuris. The upper age group (45 years above) shows only a slight increase. Among the Kabuis, there appears to be a gradual rise of total pregnancy value from the lower to upper age group. The Tangkhul Nagas show steady increase value of pregnancy rate from the lower to upper age group. The Muslims follow very closely the Tangkhul Nagas in respect of conception, survival, child loss and abortion rate. There is practically no male-female sex-ratio difference among the four groups (Table 2.4). The chi-square test on living vs. dead children reveals significant values ($P > 0.001$) especially between the Meitei and Tangkhul; the Kabui and Tangkhul; the Tangkhul and the Muslims (Table 2.5). The mean fertility rate is more among the Muslims (while the three other groups behave almost uniformly.

The Kabui Nagas record the highest mean age at menarche (15.15 years) while the Muslims have the lowest values (10.93 yrs). The t-test yields significant difference between the Meiteis and Muslims. (Table 2.8). The Manipuri Muslims as contrast to Assam Muslims, have a very early menarcheal age. The importance of the estimation of age, socio-economic factors affecting the menarcheal age have been stressed. (Table 2.9). The three valley population groups show a gradual increase of mean fertility values alongwith the increase of the I value. The almost similar trend clearly supports Spuhler (1962) indicating that I remains relatively high in city and town populations. The lower value of I among the Tangkhul Nagas (hill) is very interesting. (Table 2.11). The lower rate of premature deaths among the Tangkhuls which may be due to their improved living conditions supports the earlier studies. The overall-results on selection intensity, however, do support Crow (1958) indicating that I increases with the decrease of the family size.

II. ANTHROPOMETRIC STUDIES

The Meitei Manipuris are characterized by short stature, mesorrhine, mesocephalic and leptoprosopic face. When compared with the only available data of this population group, it indicates that within a span of 80 years, there is a decrease of about 35 mm or about an inch in the total body height of the

Manipuris.

The Muslims of Manipur, on the other hand, show a difference of almost 100 mm when the present data is compared with the available Muslim data from Bangladesh and North Bengal. Compared to the Muslims of North Bengal and Bangladesh, the Muslims of Manipur has very short stature in both sexes. Both sexes are characterized by long faces, mesorrhine, dolicho-mesocephalic head forms. In the absence of any available data, on this group from the state, we are unable to assess properly the ethnic affiliation. The Tangkhul Nagas living in the hills are characterized by below-medium stature while the Kabuis living in the valley are found with short stature. Although, both are having mesorrhine form of nose yet they vary in head shape and form. The Vangkhuls possess medium head while the Kabuis have long and narrow head. Probably this is genetic in nature. The Kabui Nagas of Manipur completely agrees with that of Haddon's (1924) classification. The Tangkhuls, on the other hand, can conveniently be grouped together with the Nagaland Nagas in the morphometric characters.

III. DERMATOGLYPHICS

Chakravartti (1963) proposed three sets of whorl-loop ratios in respect of finger prints among the mongoloid populations of North-East India. The digital patterns of the two Manipuri groups and the three Manipuri Naga groups pose a general question. In all the five groups, the whorl-loop occurrence presents uniformly in both sexes excepting the Meitei females. They also do not follow the male pattern like other groups. The similar trend of the Meitei females among the Kabuis as well as that of the Bishnupriya group is very interesting and needs further probe. The Mongoloids and the Mundaris appear to stand close to one another in respect of papillary patterns than either of them to the Australoids. So, it is not unlikely that a few Mongoloid population groups of North-East India have certain Mundari-speaking digital patterns. In Manipur, therefore, all the three sets of whorl-loop ratios are present alongwith a new set having close similarity with the formula of the Mundari-speakers. So far the palmar main line formulae are concerned, the Meiteis present different modal type for the three main line formula of D, C, B and A in the two sexes. The males show 7.5.5.>9.7.5.>1.9.7. while the females have 7.5.5.>11.9.7.>9.7.5. This was unique with the Meiteis only. The three Manipuri Nagas both sexes follow the same modal type like that of the Meitei males. Among the Kuki-Chin groups, some deviation is noticed. Both the Thadou and the Veipei Kukis follow the palmar formula of the Manipuri Nagas. The Lushais show 9.7.5.>7.5.5>,11.9.7. It may, therefore, be said that the Manipur population groups show the highest frequency of 7.5.5. main line formula followed by 9.7.5. and 11.9.7.

In a very specialized study, Chakravartti and Chakravartti (1980) did not observe any significant difference between the patients and controls in relation to the finger and palm prints of duodenal and peptic ulcers.

IV. BLOOD GROUPS

When the ABO distribution of the four Manipur population groups are compared, it appears that the Meiteis and the Muslims, the two religious communities of the valley have the same blood group picture (O>A>B). The Kabui follow the valley population while the Tangkhuls show A>O>B. This pattern is, however, closely agree with that of the Tangkhuls. The inter-group homogeneity values, however, show significant results between Meitei X Kabui and Tangkhul X Kabui. The frequency of the gene r has been found to be high in all the four groups studied. The next in order of frequency is the p gene.

Among the two Naga groups from the Manipur Hills, the gene N is rather high in the two groups compared to other populations of Nagaland. The Nocte Nagas (Bhattacharjee, 1957) show lower N genes (0.198). The two communities from the valley have almost similar distribution of MN blood groups. The chi-square test reveals a satisfactory agreement between the observed and expected number of phenotypes.

R_1 gene appears to occur in highest frequencies in all the four studied groups. R_2 genes occur in second highest frequency. Among the Kabuis, the gene Rz is absent, while among the Tangkhuls and Meiteis, this gene occurs in equal frequencies. The Muslims have the highest frequency of this gene. The gene r (cde) is absent among the Kabui and the Muslims while the Tangkhul and Meitei show this with low frequeucy. A review of the ABO blood group distribution among the population of North-East India been attempted. Similarly, the ABO distribution of the plains populations of North East India has been discussed. The Mongoloid element which is dominant in almost all the population groups of Assam, Nagaland and Manipur has under gone change to great extent. This has, however, not disturbed the ABO blood group frequency of the Muslims of Assam as well as of Manipur. The Muslims of Manipur behave uniformly with those of other local population groups.

The close similarity in ABO blood groups between the Tangkhul Nagas and the Meiteis is again confirmed indicating the ethnic homogeneity between these two groups. This was previously observed in respect of fertility behaviour, anthropometry and dermatoglyphics.

V. V. HAEMOGLOBIN E GENE

While the high prevalence of Hb E is a characteristic pattern in Assam, this gene has a low incidence in Manipur. The variation in the frequency of this gene in Assam and Meghalaya may be attributed to the degree of admixture, while in Manipur the Meiteis have not allowed so much gene flow from Assam or Meghalaya. The admixture was more pronounced with the local tribals and in the border areas of Burma. Continuation of the search for this gene among the local Naga groups of Manipur, Nagas from Nagaland, and the border areas of Burma may help in determining the degree of admixture.

VI. FUNCTIONAL ASSYMMETRY

In respect of Hand Clasping, the Meiteis, Tangkhuls and the Kabuis follow the similar pattern of R>L while the Muslim show L>R.

Similarly, all the above three groups show L>R in respect of Arm Folding against R>L in the case of the Muslims. Significant inter-group differences are observed between Meitei and Tangkhul and Meitei and Muslim, so far hand clasping is concerned. In respect of arm folding there is inter-group difference which is significant at 1 per cent level between Kabui and Muslim. In other two traits, there exists no inter-group differences. When the three combinations hand clasping, Arm folding and sitting posture are considered together, there exists significant inter-group differences between Meitei and Kabui; Meitei and Muslim and Tangkhul and Kabui Nagas. The Tangkhul and Meitei show significance at 1 per cent level. The same picture emerges between Tangkhul and Muslim.

The frequencies of these three combinations have been discussed in Table 7.5 The present data on hand clasping and arm folding from the Meiteis agree with that of Singh and Malhotra (1971).

VII. EAR LOBE AND TONGUE-ROLLING

There exists no sex difference in respect of ear lobe attachment, in the four groups studied. The Kabui X Muslim; Tangkhul X Muslim and Meitei X Muslim show significant inter-group difference. The importance of this trait (which is based on polygenes) as a taxonomic character has been emphasized. There is no difference in the occurrence and distribution of the tongue-rolling in the four groups.

VIII. COLOUR BLINDNESS AND TASTE SENSITIVITY

The Tangkhul Nagas show the lowest incidence of colour blindness (2.66) while the remaining three groups show almost equal incidence. There are no inter-group differences. The incidence of colour blindness among the various reported population groups of North-East India has been discussed.

The t gene varies from 36.95 to 60.46 per cent. Excepting the Muslims the t gene in the other three groups occur in a similar frequency. The inter-group differences do not exhibit any statistically significant value. In the absence of any available Manipur data, the present data have been compared with those of North-East India.

IX. CERUMEN TYPES

The genetic dimorphism and distribution of human cerumen types suggest that dry allele is significantly predominant over that of the wet allele. It appears that the high incidence of dry allele is the characteristic of all the population groups of North India studied so far. The Kukis occupying the South-West of Manipur have a comparatively low value of dry allele than those of the Nagas as well as those of the Manipuris, but presenting a genetic homogeneity among themselves. The two Manipuris from the central valley have again very close similarity not only between them but also with those of the Manipuri and Nagaland Nagas.

In Assam, the caucasoid castes and the mongoloid tribes live side by side and in course of time, exchanges of gene took place between the tribals and the castes. Even then a gradient could be drawn in respect of their dry genes.

In Manipur, the admixture was more with the local neighbourhood Nagas that is probably why the genetic cline could be evident.

Acknowledgement

In the preparation of this chapter, free use has been made of the articles by B. M. Das of the Anthropology Department, Gauhati, Assam. We gratefully acknowledge our gratitude to him and to his team of researchers who have helped us with their literatures on North-East India especially of Assam.

Reference

1. Basu, R.N. (1938) : "Blood Groups among the Khasi", *Nature*, 142 : 797.
2. Bhattacharjee, P.N. (1954) : "Blood group investigation in the Abor tribe", *Bull Dept. Anth. Govt. of India*, III : 1.

3. ———. (1955) : cited by B.M. Das (1974), in *Contemporary Anthropological Research in North East India.*
4. ———. (1957) : "A study on ABO, MN blood groups and the ABH secretion in the Nocte of NEFA", *Bull. Dept. Anth. Govt. India*, VI: 1.
5. ———. (1949) : "ABO blood groups of the Angami Naga", *Eastern Anthropol.* II : 1.
6. Bowles, GT. (1943) ; Linguistic and racial aspects of the Munda problem," *Mem. Peabody Mus. Paper Anth. Arch. Eth.* XX.
7. British Research Association (1939) : "Blood groups among primitive peoples," *Nature*, 144 : 714.
8. Brown, J.C. and Kemp, S.W. (1915) : "Abors and Galongs by Dunbar", *Mem. As. Soc. Beng.* Part V.
9. Buxton, L.H.D. (1925) : *The Peoples of Asia*, London.
10. Chakravartti, M.R. (1963) : "Ethnic bearing of the dermatoglyphic studies in India." *Acta Genet Med.* Gemellol (Roma) XII : 4.
11. ———. (1963) & D.P. Mukherjee (1961) : "*Dermatoglyphics of the* tribes and castes of Assam", *Science & Culture* 27 : 2.
12. ———. and Rama Chakravartti (1980) : *Les Dermatoglyphes et les ulceres du duodenum et de L-estomac L' Anthropologie* : 80 : 126-128.
13. Crow, J.F. (1958) : "Some possibilities for measuring selection intensities in man", *Hum. Biol.* 30 : 1-13.
14. Das, T.C. (1945) : *The Purum*, Calcutta.
15. Das, A.K. & S.K. Banerjee (1972) : "Finger prints of the Ahoms of Assam," *Bull. Anth. Deptt.* Dibrugarh University. I.
16. ———. (1973) : "Palm prints of the Ahom,' *Bull. Deptt. Anth.* Dibrugarh University. II.
17. Das, P.B. & B.M. Das, (1967) : "Age at menarche of Kalita girls of Assam", *Man in India*, 47 : 2.
18. Das, B.M. (1960) : *Ethnic affinities of the Rabhas*, Calcutta.
19. ———. (1960) : "Somatic variations among the Hills and Plains Garos of Assam," *Man in India*, 40 : 2.
20. ———. (1960) : "Ethnic elements in some Bodo populations of Assam," *Jr. As. Soc. Res. Soc.* XIV.
21. ———. (1964) : "Somatic variability among some populations of South Goalpara, Assam," *Proc. Nat. Inst. Sc. India*, 30 : 5-6.
22. ———. (1968) : *ABO blood groups in the tribal populations of North East India with special reference to the Khasi*, Anthropologist (special) vol.
23. ———. (1971) : "Taste sensitivity to phenylthiourea among the Khasi of North East India," *Proc. Int. Symp. on Hum. Genetics, A.U. Waltair*, Ed. M.R. Chakravartti (1973).
24. ———. (1974) : "*Contemporary anthropological research in North East India*," Ed. B.M. Das, Dibrugarh University, Dibrugarh, Assam.
25. ———. (1970) : "Somatic variability among the Khasi populations of Assam India," *Z. Morph. Anth.* 62 : 3.
26. ———. (1958) "Demography of a Rabha village," *Vanyajati*, VI : 2.

27. ———. andMahapatra, U.D. (1958): "Finger prints of the Suri", *J. Gau. Univ*, IX.
28. ———. (1960) : "Physical characters of the Suri," *J. Gauhati University*, XI.
29. Das, B.M. and R. Das (1965) : "An Anthropometric Study of the Kalita," *Man in India*, 45 : 2.
30. ———. and A.C. Bhagabati (1962) : "Investigations on the finger prints of the Khamyang," *J. Gau. Univ*. XIII : 1.
31. ———. (1963) : "Somatometric characters of the Khamyang," *J. Gau. Univ*. XIV : 2.
32. Datta, B. (1938) : "Anthropological notes on some Assam castes," *Anth. Papers*, Calcutta. Univ. 5.
33. Dekaz R. (1970) : *cited by B.M. Das*. (1974).
34. Deka, R. and A. Bora (1973) : "An investigation on the finger prints of the Mikirs," *Bull. Anth. Dept. Dib*. Univ. II.
35. Dixon, R.B. (1921) : "The Khasis and the racial history of Assam," *Man in India*, 2.
36. Dutta, D. (1973) : "Finger prints of the Singpho females," *Bull Dept. Anth*. Dibrugarh University II.
37. Flatz, G. et. al. (1972) : "Gentic survey in the population of Assam I, ABO blood groups, Hb and G_6 PD type," *Human Heredity* 22.
38. Goedde, H.W. et. al. (1972) ; "Genetic survey in the populations of II serum protein and red cell enzyme polymorphisms," *Human Heredity*, 22.
39. Gogoi, D. (1972) : "Menarche and monopause among women of an Ahom village of upper Assam," *Bull Dept. Anth*. Dib. Univ. I.
40. Guha, B.S. (1931) : "The racial affinities of the peoples of India," *Census of India* I : 3.
41. ———. and P.C. Basu (1931) : "A report on the human relics recovered by the Naga hills expedition for the abolition of human sacrifices during 1926-27," *Anth. Bull. Zoo. Surv. Ind*. I.
42. Gupta, S.P. (1958) : "ABO blood groups in Tripura (North East India." *J. Royal Anth. Inst*. 88.
43. Haddon, A.C. (1924) : *Races of Man*, Cambridge, London.
44. Hutton, J.H. (1921) : *Primitive tribes in Modern India and the West*, London.
45. Kumar, N. (1954) : "Blood group and secretor frequency among the Galong," *Bull. Anth. Dept*. Govt. India III : 1.
46. ———. (1958) : "ABO blood groups and sector factor distribution among the Noatia, a mongoloid tribe of Tripura state," *Anthropol*. V: 1-2.
47. ———. and D.B. Sastry (1961) : "A genetic survey among the Riang, a mongoloid tribe of Tripura," *Z. Morph. Anth*. 51 : 3.
48. Macfarlane, E.W.E. (1941) : "Tibetan and Bhotoa blood group distribution," *J.A.S.B*. 7.
49. Majumder, D.N. (1950) : "ABO blood groups among the Garos of

Eastern Pakistan," *Man in India*, 30.

50. Miki et al. (1960) : "On the distribution of secretor charaeter of the Lepchas and the Khasis," *Proc. Jap. Acad.* 36 : 4.

51. ———. (1960) : "Investigations on the finger prints of the Lepchas and Khasis," *Jap. Acad. Proc.* 36 : 6.

52. ———. (1960) : "On the distribution of the ABO blood groups and the taste ability for PTC of the Lepchas and the Khasis," *Jap. Acad. Proc.* 36 : 2.

53. Mitra, P.N. (1936) ; "Blood groups of the Angami Naga and Lushai tribes," *Ind. J. Med. Res.* 63.

54. Olivier, G. (1958) : "Physical Anthropology of the Nagas of Assam", *Man in India*, 38 : 2.

55. Pakrasi, K. (1953): "A study of physical characters of the Arleng" (Mikir), *J. Gau, Univ.* IV.

56. Phookan, M.N. (1961) : "Anthropometric characters of the Jharua Kachari of Assam", *J. Gau. Univ.* XII : 2.

57. ———. (1962): "Anthropometric characters of the Sonowal Kachari of Assam", *Bull. T.C. & F.R. Gau. Univ.*

58. Phookan, M.N. (1974) : *Anthropometric characters of the relations in respect of Anthropometry, ABO blood groups and taste sensitivity,* unpublished Ph.D. thesis of Dibrugarh University, Assam.

59. Rakshit, S. (1960): "A short note on the menarcheal age of Assamese girls", *Man in India,* 40.

60. Raychaudhuri, T.C. (1935): "The Khasis", *J. Dept. Lett.*, Cal. Univ. 26.

61. Sharma, T.C. (1962): "Anthropometry of the Deuri", *J. Gau. Univ.* XIII : 2.

62. ———. (1961): "Physical characters of the Miri", *J. Gau. Univ.* XIII : 2.

63. Shaw, W. (1928): "Notes on the Thado Kuki", *J.A.S.B.* 24.

64. Singh, N.R. and K.C. Malhotra (1970): "Ear lobe among the Brahmins and Meiteis of Manipur", *Bull. Deccan Coll. Res. Inst.* XXIX : 165-168.

65. ———. (1971): "Hand Clasping and Arm Folding among the Manipuris (India)", *Hum. Hered,* 21 : 203-207.

66. Singh, S. (1935); "A contribution to Angami and Sema somatology", *J A.S.B.* I.

67. Shrivastava, R.P. (1969): "Finger Prints, colour blindness and other features in the Miris of Assam", *Man in India,* 49 : 4 : 398-403.

68. Spuhler, J. (1962); "Empirical Studies on Quantitative Human Genetics", in *Proceedings of UN/WHO seminar on the use of vital and health statistics for Genetics and Radiation studies,* United Nations, New York pp. 241-252.

69. Waddel, L.A. (1900): "The tribes of Brahmaputra valley", Sanskaran Prakashak, Delhi (1975). *J.A.S.B.* LXIX.

Index